OTHER TITLES OF INTEREST FROM ST. LUCIE PRESS

Ecological Integrity and the Management of Ecosystems

Economics of the Tropical Timber Trade

Emergency Response Manual for Chemical and Hazardous Material Release

Environmental Effects of Mining

Environmental Fate and Effects of Bleached Pulp Mill Effluents

Everglades: The Ecosystem and Its Restoration

Family Agriculture

From the Forest to the Sea: The Ecology of Wood in Streams, Rivers, Estuaries, and Oceans

Sustainable Agricultural Systems

Sustainable Forestry: Philosophy, Science, and Economics

The Everglades Handbook: Understanding the Ecosystems

Tropical Forestry

For more information about these titles simply call, fax or write St. Lucie Press.

St. Lucie Press
100 E. Linton Blvd., Suite 403B
Delray Beach, FL 33483
TEL (407) 274-9906
FAX (407) 274-9927

Economic Theory for Environmentalists

Economic Theory
FOR Environmentalists

John Gowdy
Department of Economics
Rensselaer Polytechnic Institute
Troy, New York

Sabine O'Hara
Department of Economics
Rensselaer Polytechnic Institute
Troy, New York

SOIL
AND WATER
CONSERVATION
SOCIETY

$S{\overset{t}{L}}$

St. Lucie Press
Boca Raton   Boston   London   New York   Washington, D.C.

Library of Congress Cataloging-in-Publication Data

Gowdy, John
    Economic theory for environmentalists / by John Gowdy and Sabine O'Hara
      p. cm.–(Total quality series)
    Includes bibliographical references and index.
    ISBN 1-57444-003-9 (alk. paper)
    1. Industrial procurement–Management. 2. Total quality
management. I. Title. II. Series.
HD39.5.F46 1995
658.7'2—dc20   94-9069
                                                       CIP

CONTENTS

FOREWORD

I write these words as an environmentalist, likely addressing others of the same stripe. To us, I think, economists sometimes seem like deaf people in a burning cinema, still watching the pretty pictures as the smoke alarm sends everyone else stampeding. This, therefore, is a powerful book, one that should have been written long ago, for it makes clear to the rest of us just why those pretty pictures are so arresting. It is the neat logic, the sheer elegance, of neoclassical theory that makes it so appealing—and in some ways so dangerous, at least in a world that seems to be reaching certain physical limits.

Some may find it frustrating to have to learn at least a little bit of the language and logic of the economist. But for better *and* for worse, it is the *lingua franca* of this age, the tongue most widely spoken, the religion most devoutly worshipped. To be a part of the debate, you need to understand the assumptions of neoclassical economics, in the same way that all philosophers of medieval Europe were Christian philosophers, or that every thought in certain barrooms is expressed in terms of the Knicks. To understand efficiency and optimality as economists understand them is to understand much about how our world works, a useful thing not only for environmentalists but for anyone concerned with or curious about the impact humans have on the world around us. This is not to say, necessarily, that you must accept each brick in the neoclassical edifice. As the authors make marvelously clear, some of those assumptions are shakier than they used to be. The scarcity, for instance, of extra atmospheres into which to pour the exhalations of our economy will challenge economists in profound ways in the years to come. In fact, the next great human debate may be between those who still desire an ever-expanding economy and those who are coming to fear its effects.

But to engage in that controversy, or more likely to shape the smaller twists and turns in public policy on issues from groundwater pollution to agricultural subsidy, you need a working knowledge of economic assumption and belief. The authors have provided that blueprint, and they have helpfully labeled some of the defects in the design. Their straightforward and unhysterical presentation is precisely what we need most. And if from time to time you are dismayed by the graphs and charts, comfort yourself with the thought of all the economists who will soon be forced to understand chemistry, physics, and biology.

Bill McKibben

Bill McKibben is an environmental writer who has published and lectured widely. His best selling books are *The End of Nature* and *The Age of Missing Information*.

ACKNOWLEDGMENTS

A book is a collaborative project, and this one is no exception. Far beyond our own collaboration as authors, there are many whose contributions have helped shape this book. First and foremost, these contributions have come from our students, whose comments, probing questions and sometimes sheer amazement at the worldview of mainline economists planted the seed for this project. Since many of our students are majoring in disciplines other than economics, their comments have broadened our own disciplinary perspective and challenged us to clarify familiar concepts. Peg, Chuck, and Valerie in particular have given us many helpful comments.

Second, thanks are due to our colleagues whose writings and personal conversations have influenced our thinking. More specifically, Douglas Booth, Steve Breyman, Terry Curran, Carl McDaniel, Dick Shirey, and Jean Stern have provided many helpful comments on earlier drafts of this book.

A third group to whom thanks are due are the many committed people we have met over the years whose work in environmental organizations, community groups, the private sector, and government agencies has advanced the idea of an economics text that would contribute to the broader dialogue necessary to understand the complex world in which we live.

Our editor, Sandra Koskoff, deserves our thanks for her tireless efforts in moving us along and keeping up her spirits in the process.

And last but not least, thanks are due to our families, particularly to our spouses. Without Linda's efforts in drafting the numerous graphs and Phil's repeated proofreading, we would not have completed this book. Thanks to you all.

John Gowdy

Sabine O'Hara

February, 1995

"The purpose of studying economics is not to acquire a set of ready made answers to economic questions, but to learn how to avoid being deceived by economists."

—Joan Robinson

(quoted in John K. Galbraith, *Economics and the Public Purpose.* Boston, Houghton-Mifflin, 1973)

1 INTRODUCTION

The most serious problem our civilization faces is the on-going conflict between economic activity and the biological world upon which all human activity ultimately depends. The purpose of this book is to explain the basic theoretical framework most economists use to describe economic activity and the relationship between this activity and the natural world. This theory is called *neoclassical economics*. Understanding the logical apparatus of this theory is important for two reasons. First, neoclassical economics shows why market forces and biological integrity are often in conflict. Second, this theory dominates the environmental policy debate, particularly in the United States. Those concerned with policy and evaluation questions should understand the basic assumptions and theoretical framework of this theory.

Neoclassical economics is a theory with a centuries-long history, a central core of commonly-held assumptions, and an elaborate mathematical scaffolding, features which have made economics the "Queen of the Social Sciences." It is called "neoclassical" because it is a mathematical elaboration and refinement of the ideas of the classical economists, whose ranks include Adam Smith

(1723–1790), David Ricardo (1772–1823), and Thomas Malthus (1766–1834). The term neoclassical was first used in the year 1900 by the great American economist and social critic Thorstein Veblen (1857–1929), who saw a continuity between the behavioral assumptions of classical economics and modern economics.

Another term for neoclassical economics is *marginal* economics. Beginning with the *marginalist revolution* in the 1870s, economists began to explain economic phenomena in terms of very small (marginal) changes around an equilibrium point, by applying the tools of differential calculus. For example, the notion of marginal utility seeks to explain how a small change in the consumption of one good affects a consumer's utility, or level of satisfaction. The use of marginal analysis in economics can be traced to a number of economists who first cast economic theory in the language of modern mathematics. The great synthesizer of classical and marginalist economics was the British economist Alfred Marshall (1842–1924).

After a temporary setback as a result of the Great Depression and theoretical assaults by John Maynard Keynes (1883–1946) and others, neoclassical economics flourished after World War Two. During the last forty years neoclassical economics has incorporated many, but by no means all, of Keynes' critiques of classical theory into what Paul Samuelson refers to as the *neoclassical synthesis*. By the 1960s Samuelson, the greatest contemporary interpreter of neoclassical economic theory, could remark that only economists of the extreme right or extreme left were not neoclassical. The dominance of neoclassical theory began to break down with the energy crisis and stagflation of the 1970s and slow economic growth in the 1980s, but it still represents the mainstream of economic thought.

Economic theory is divided into *microeconomics*, whose subject matter is the individual decision-making units in the economy, the firm and the consumer, and *macroeconomics*, which is concerned with broad aggregates of economic activity such as unemployment, inflation, and economic growth. This book deals mainly with microeconomic theory. The reason for this focus is that microeconomics is the basis for most environmental policy recommendations of economists. It also provides the foundation for several schools of macroeconomic thought, particularly those promoting market-based environmental policy recommendations. A survey of the various schools of macroeconomics is presented more completely in Chapter 7.

MARKETS AND MODELS: THE CIRCULAR FLOW OF ECONOMIC ACTIVITY

Neoclassical theory as presented in introductory economics textbooks describes economic activity as a self-contained and self-perpetuating circular flow between producers and consumers, the two basic categories of participants in the economy. Households (consumers) provide firms (producers) with labor and other productive inputs, and firms provide households with goods and services (see Figure 1.1).

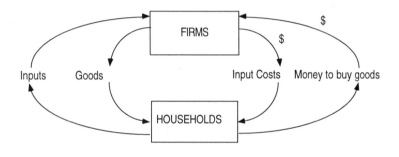

Figure 1.1   The Circular Flow of Economic Activity.

Through the institution of *markets*, goods and inputs are traded among these economic actors. In the examples in the chapters that follow, a market will be considered a particular location where exchange takes place. In standard economic theory markets are more broadly defined as "mechanisms" to facilitate exchange. The market that involves the trading of finished products among consumers is called the *goods and service* or *output* market, and the market that involves the trading of factors of production among firms is called the *factor* or *input* market. In the goods and service market the outputs traded include everything from food, clothing, and cars to the services provided by the real estate, finance, and insurance sectors.

About two-thirds of the total output of goods and services in the economy is used to produce other goods and services. These are called *intermediate products* and include such things as the glass produced to make automobile windshields or the fabric to make clothes. About one-third of the total output of the economy represents *final products* sold directly to consumers, such as automobiles, food and clothing, or a haircut at the barber shop. When economists calculate *gross national product* (GNP) as a measure of all the output produced, only those goods and services are counted that are sold to final users. Otherwise, economic output would be overestimated by counting intermediate products twice—once, when they are first produced by one firm and again as part of the final product of another firm. For example, the value of windshields produced would be counted once as an output of the glass industry and then again as an output of the automobile industry. Those goods and services that are not sold in markets, and for which no money is exchanged, are not counted at all. They do not show up in the traditional GNP accounts.

Goods and services are produced using productive inputs or *factors of production*. Factors of production are classified by economists into three broad categories: *land*, *labor*, and *capital*. Sometimes, a fourth category, *entrepreneurship*, is added. It refers to the contribution of entrepreneurial organization and leadership to the production process. The factor *land* includes not only acres of land but all natural resources. All materials not modified by humans, including raw materials such as iron or copper and primary energy such as petroleum or coal, are counted as "land." *Labor*, as the name implies, is the input of human labor power (measured in worker hours or number of persons employed) necessary to produce outputs. The term *capital*, as used by economists, does not refer to money but to all the machines, tools, buildings and structures, and other human-made artifacts used to produce goods and services.

In the process of producing goods and services, these factors of production, land, labor, capital, and entrepreneurship earn income in the form of, respectively, rent, wages, interest, and profit. The income creation side of economic activity is captured in the *value-added* portion of the national income and product accounts. Value-added is the additional economic value created at each step of the production process, for example, the additional value created when beef is transformed into hamburgers.

Neoclassical economics is a model of how markets work to distribute given amounts of goods and services. As any scientific model, it is an abstract representation of reality. A good model makes as few simplifying assumptions as possible to capture the essential features of the reality it describes. The simplification of neoclassical theory is to focus only on the efficiency with which given amounts of resources are used to produce goods, and the efficiency with which these goods are distributed among consumers. It is not concerned with where market goods ultimately come from or where they end up after they leave the economic system. As many critics of neoclassical theory have pointed out, it contains no concept of scale. No matter how large or small the economy is in relation to the rest of the world (in terms of available natural resources or the assimilative capacity of environmental sinks such as the oceans or the atmosphere), there is always a single "most efficient" allocation of available resources. The lack of consideration of scale or of a reality other than market exchange is a serious flaw in the model, because the scale of economic activity is, in fact, critical in a finite biophysical world.

ECONOMIC EFFICIENCY AND PARETO OPTIMALITY

The concept of efficiency is central to neoclassical economic theory, and a major purpose of this book is to explain exactly what economists mean by this term. Efficiency has a very narrow meaning in neoclassical theory. The analysis of efficiency in consumption begins with some fixed amount of the goods to be traded and some initial distribution of these goods among consumers, much like in an auction. The fairness of the initial distribution, the ultimate source of consumer goods, and their ultimate destination as waste released into the biophysical environment, are subjects outside the scope of this theory. In a similar fashion, the neoclassical analysis of efficiency in production begins with a given distribution of productive inputs and a given technology. Where these inputs ultimately originate is outside the focus of the analysis.

Ideally, the end result of unhindered economic exchange is *Pareto optimality*, a concept central to neoclassical theory. The concept gets its name from the fact that it was first formally stated by the Italian economist and social philosopher, Vilfredo Pareto (1848–1923). When Pareto optimality is achieved in the goods market, no further trading of consumer goods can make one individual better off without making another individual worse off. Pareto optimality in production means that no further trading of inputs between firms can increase the production of one good without decreasing the production of another good. This condition is what neoclassical economics means by efficiency.

Pareto optimality is the end result of a successful trading process of goods and services as well as inputs. Guided by Adam Smith's "invisible hand," individual consumers and producers follow their own self interest to bring about the best for society as a whole. But what is meant by "best for society as a whole"? Smith believed that there is no conflict between an individual and a social optimum. But he also made some strong assumptions upon which he based his theory of the effectiveness of the invisible hand. One of Smith's assumptions was that people have a strong sense of moral obligation and responsibility that sets the stage for market exchange. This is in contrast to neoclassical economics, which assumes "value neutral" economic actors.

Neoclassical theory focuses only on what happens inside the sphere of market exchange. It is a theory of allocation, that is, a theory dealing with the most efficient distribution of scarce resources among the various ways in which they can be used. Although neoclassical economics is sometimes called "price theory," prices in the neoclassical world have only a limited purpose. They exist merely to facilitate the exchange of goods, services, and factor inputs in an economy too complicated to operate as a barter system. Money is a means of exchange making it possible to trade various "unequal" goods, and it is used to assign a commonly accepted exchange value to them.

THE CONTEXT OF MARKET EXCHANGE

Neoclassical theory is indispensable in describing of the power of the market as an institution for allocating given amounts of goods among consumers and inputs among producers. It says little, however, about how the economic activities of producers and consumers affect the stocks of natural resources or the quality of the environmental media that receive wastes and emissions. Since only market goods and inputs are part of neoclassical theory, most social and biophysical relationships lie outside the scope of the theory. These relationships include non-market economic activities such as subsistence production or domestic contributions, other human activities not concerned with production or consumption, and finally, the biophysical world within which humans live. All these relationships are depicted in Figure 1.2.

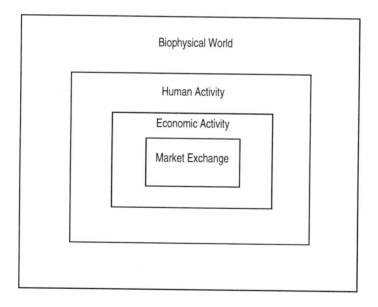

Figure 1.2 Economy-Environment Interactions.

All *human activity* on the planet takes place in the context of the *biophysical world*, which includes all the biological, atmospheric, geological, and chemical processes that make up planet Earth.

Human activity includes all human effects on the natural world as well as the social relationships among humans. The term also includes biophysical processes, such as releasing energy from physical activity, breathing, decaying and so on, and all economic activity. Human activity has grown so large that it threatens to disrupt natural cycles and processes which have evolved over eons. The human population is now over 5.6 billion and, according to mathematical calculations, is almost certain to double by the middle of the next century. According to some estimates, the human species now appropriates or co-opts about 40 percent of the potential terrestrial, net primary biological production of the planet. Net primary biological production is a measure used to describe the productivity of plants through photosynthesis. It refers to the solar energy green plants convert via photosynthesis into organic matter minus the energy plants use for their own life processes.

Economic activity may be described as the day-to-day consumption and production activity that humans engage in to provide themselves with the material and nonmaterial requirements of their existence. Within the category labeled economic activity is *market exchange*. Market exchange is the only part of human economic activity described by neoclassical economic theory. It is an important part of economic activity but is certainly not the only one. Many other economic interactions take place outside of markets, in home gardens, families, communities, in the so-called "informal sector," in barter exchange and many other non-measurable, non-market kinds of economic interaction. Market exchange is only a part of this total economic interaction and an even smaller part of total human activity, or of the entire planetary activity labeled *biophysical world*. The theory of market exchange ignores processes taking place in the larger biophysical world, the social context of economic behavior, and any other economic activity that does not involve directly measurable market transactions. While it would be useful to examine *all* the connections and interactions between the different levels of contexts and interactions shown in Figure 1.2, this book's critique of neoclassical theory will focus on the relationship between market exchange and the biophysical world within which all human activity takes place.

In describing market exchange, neoclassical theory performs a valuable service. While the area of market exchange at the center of

Figure 1.2 may be small, it is most influential. Not only does it impact the levels of activity in Figure 1.2 outside market activity, it also shapes how humans see the rest of the world. The preoccupation of contemporary social and political institutions with market exchange drives the public policy agenda and influences the ways in which society defines the functions and services of the biological world.

ECONOMICS AND THE BIOPHYSICAL WORLD

The economy interacts with the biophysical world of energy and material flows, and biological processes, in two major ways: as a source of raw materials and as a sink for the waste produced by economic activity. In addition, market exchange determines how such things as water, air, forests, soil, whole ecosystems, and even characteristics of the atmosphere are valued in a modern economy. In fact, all these are commonly summarized as "the environment," a term which makes no valuation distinctions according to ecological functions or contributions. This is a good example of the homogeneity neoclassical economics assumes.

The fact that the origins of consumption and production are ignored is significant. The basic model of market exchange is abstracted from time, place, and social and environmental context. Economic markets are merely the meeting places of producers and consumers stripped of all history, social context, and biophysical reality. Place is reduced to transportation costs, and time to a single point—the immediate present.

WHAT THIS BOOK DOES AND DOES NOT DO

Within its limited context, neoclassical theory is, in many ways, an accurate description of modern market exchange. The environmental crises we face today are fundamentally the result of forces described eloquently by neoclassical theory. For this reason, it is essential that those concerned with the health of the planet understand the concepts behind the economic theory that to a great extent shapes decisions affecting our home—Earth. How are "economic" values for the components of the natural world calculated?

When are market criteria appropriate for the allocation of these components? What are the limitations of the neoclassical worldview? When should the market be amended or overridden?

These are the questions addressed by this book. Its purpose is to present the basic concepts of neoclassical theory to persons interested in the implications of this theory for policies affecting the natural world. This discussion of economic theory is limited to the three basic building blocks of neoclassical economics:

1. the notion of Pareto optimality and the theory of general equilibrium;

2. the model of perfect competition, which implies that free markets will achieve Pareto optimality; and

3. market failure—the idea that if incorrect price signals are sent, market economies will fail to achieve Pareto optimality.

These building blocks of microeconomics and the relevance of the assumptions underlying them are presented in Chapters 2–6. Chapter 5 also includes some additional economic concepts useful for policy analysis such as price and income elasticities, consumer surplus, and some market-based methods for measuring the economic value of environmental resources. Throughout these chapters the example of two goods—beef and Brazil nuts—are used to discuss the implications of neoclassical theory for environmental policy. Chapter 7 covers a brief history of economic thought, focusing on the connection between microeconomic theory as described in this book and various schools of macroeconomics. It also introduces a new field of study that seeks to explicitly link economic and ecological concepts—ecological economics. Chapter 8 examines a range of policy options and valuation concepts as they apply to the specific problem of groundwater pollution. This example illustrates how theoretical assumptions and concepts ultimately determine environmental policy and shape our view of the value and usefulness of natural resources.

Economics has a long and rich history, much of which is ignored in this book. Models of market structure other than perfect competition, as well as the market for production inputs only receive a brief mention in Chapters 6 and 8 respectively. In spite of the fact that some of the difficulties discussed in the book have been recognized in the neoclassical literature, neoclassical economists almost invariably defend the validity of policies based on the simple model discussed in the first six chapters of this book.

Environmentalists are for the most part critical of neoclassical economic theory. Their criticisms focus on the unrealistic assumptions of the market model of perfect competition (discussed in Chapter 5) such as perfect information, the absence of barriers to trade, homogeneity of goods within a particular market, individual rationality, and so on. We argue that even if one accepts all the assumptions of neoclassical theory, Pareto efficient market outcomes would not by themselves ensure environmental sustainability. The conflict between economic activity and environmental quality is not merely the result of "market failure," nor of the fact that real-life market economies are not perfectly competitive. The economy-environment conflict ultimately arises from the impossibility of economic markets to place ecologically meaningful values on the functions and attributes of the biophysical world.

We do not argue that neoclassical theory itself is fatally flawed. But we do take exception to the claims of neoclassical economists that (1) efficiency in exchange (Pareto optimality) should be the major goal of economic policy, and (2) market exchange provides a sufficient valuation framework for all of social and ecological reality.

In the current political and economic climate with its almost exclusive emphasis on economic growth, any proposed policy to protect the natural world will be subjected to intense scrutiny based on calculations weighing economic costs and benefits. For those concerned with more than the material growth of the economy, namely with the quality of life, the long-term sustainability of the biosphere, and the larger human and nonhuman context of economic activity, it is essential to understand the theories and concepts behind economic cost-benefit calculations. We hope that this book can contribute to a dialogue across disciplines and professions. Such a dialogue is not only valuable in itself, it is essential to increasing our understanding of the relationship between the economy and an ever more threatened natural world.

SUGGESTIONS FOR FURTHER READING

The History of Economic Thought

Blaug, Mark. *Economic Theory in Retrospect.* 4th Edition. Cambridge Univ. Press, Cambridge, 1985.

Brue, Stanley. *The Evolution of Economic Thought.* 5th Edition. Dryden Press, Orlando, Florida, 1994.

Microeconomic Theory Texts

Ferguson, C.E. *The Neoclassical Theory of Production and Distribution.* Cambridge Univ. Press, New York, 1969.

Frank, Robert. *Microeconomics and Behavior.* 2nd Edition. McGraw Hill, New York, 1994.

The Impact of Humans on the Environment

Gordon, Anita and David Suzuki. *A Matter of Survival.* Allen and Unwin, Sydney, Australia, 1990.

Orr, David. *Ecological Literacy.* SUNY Press, Albany, New York, 1991.

Ponting, Clive. *A Green History of the World.* Penguin Books, London, 1991.

Vitousek, Peter et al. "Human Appropriation of the Products of Photosynthesis," *BioScience* 36 (1986), 368–373.

Critiques of Neoclassical Theory

Daly, Herman and John Cobb. *For the Common Good.* Beacon Press, Boston, 1989.

Georgescu-Roegen, Nicholas. *The Entropy Law and the Economic Process.* Harvard Univ. Press, Cambridge, Massachusetts, 1971.

Goodwin, Neva. *Social Economics: An Alternative Theory.* St. Martin's Press, New York, 1991.

Gowdy, John. *Coevolutionary Economics: Economy, Society and Environment.* Kluwer Academic Press, Boston, 1994.

Lutz, Mark and Kenneth Lux. *The Challenge of Humanistic Economics*. Benjamin Cummings, Menlo Park, California 1979.

Sahlins, Marshall. *Stone Age Economics*. Aldine, New York, 1972.

The best source for short explanations of specific topics in economic theory and economic history is *The New Palgrave Dictionary of Economics*, MacMillan and Company, London, 1987 (4 volumes).

2
THE THEORY OF THE
CONSUMER

INTRODUCTION

Microeconomic theory divides the world into two basic groups of actors—producers and consumers. The basic concepts and assumptions underlying the economic behavior of consumers are described in *demand* or *consumer* theory (Figure 2.1). What is our motivation as consumers to become actors in economic markets? How are our preferences registered and interpreted in these markets? What influences our decisions regarding the purchase of goods and services? These are some of the many questions raised in consumer theory. Neoclassical economics limits its analysis of consumer behavior to a theory of exchange. It is the purpose of this chapter to explain that theory.

Imagine the following scenario. Carl is given 200 coupons. One hundred of these are redeemable for one-pound packages of roasted Brazil nuts, and the other 100 for one-pound packages of hamburger beef patties. Since the coupons expire the next day, he decides to share them with his friends and neighbors. As he arbitrarily distributes the coupons, the people receiving them begin to trade with each other since some do not eat beef at all, others prefer beef to nuts, and still others prefer various combinations of both. As Carl's friends and neighbors negotiate and trade with each other, they improve their situation over and above the initial distribution of coupons.

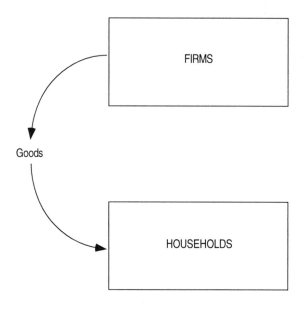

Figure 2.1 Consumer Theory.

In consumer theory the answers to questions about consumer behavior and motivation are grounded in the concept of *utility*. Consumers are motivated to participate in markets to gain utility, that is, satisfaction from the goods and services available in the market. Consumer theory describes rules of behavior that enable consumers to gain the maximum possible satisfaction from the limited amount of goods available and the limited means (income) available to acquire them. Like any theory, it begins with some key simplifying assumptions:

1. More goods are always preferred to fewer (*non-satiation*).

2. Consumers are rational and consistent in their choices (*transitivity*).

3. Consumers choose among *commodity bundles*, that is, among various combinations of available goods, generally preferring a mix of goods to having all of one kind.

4. Consumers are interested in maximizing their own utility (satisfaction) and are willing, in principle, to trade any good for any other good to achieve that goal.

THE INDIFFERENCE CURVE

The assumptions of consumer behavior are embodied in the notion of the *indifference curve*.

> **THE INDIFFERENCE CURVE SHOWS ALL THE VARIOUS COMBINATIONS OF GOODS THAT GIVE THE CONSUMER THE SAME UTILITY, THAT IS, THE SAME LEVEL OF SATISFACTION.**

Each indifference curve in Figure 2.2 shows all the combinations of two goods, X and Y (Beef and Brazil Nuts), which give a consumer the same amount of utility, or satisfaction. Consumers are willing to trade goods if this will increase, or at least maintain their level of satisfaction. Since the same level of satisfaction is achieved by consuming any combination of beef and Brazil nuts shown on the same indifference curve, the consumer is equally happy with any of these combinations. Similar to the isothermals on a weather map showing the same temperature, or contour lines on a topographical map showing the same altitude, the indifference curve might be more properly called an *iso-utility* curve—a curve whose points indicate the same level of utility. The indifference curve I in Figure 2.2 shows that consumer A is equally satisfied with 12 units of Beef (good X) and 8 units of Brazil Nuts (good Y), or 6 units of Beef and 14 units of Brazil Nuts. In neoclassical jargon, *consumer A is indifferent between these two commodity bundles.*

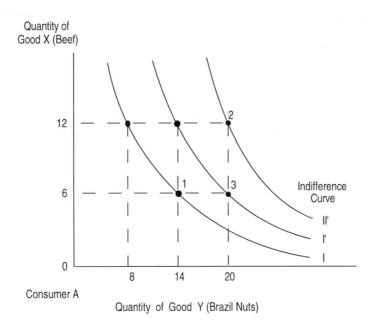

Figure 2.2 The Indifference Curve.

Although Figure 2.2 shows three indifference curves, a complete representation of a consumer's preferences would contain an entire set of indifference curves. In Figure 2.2 this can be pictured as a "consumption space" between the axes of the graph, filled with indifference curves showing all the possible combinations of goods X and Y, beef and Brazil nuts, which provide various levels of total utility or satisfaction to the consumer. Indifference curve analysis is a graphical representation of the motivation for market involvement of consumers.

The assumption of non-satiation—that there cannot be enough of a good thing, or more is always better—implies that indifference curves further away from the origin are preferred to the ones closer to the origin of the graph. In Figure 2.2 imagine a straight line drawn from the origin to any point between the two axes of the graph, with indifference curves crossing it. The further way from the origin an indifference curve is, the more of both goods the consumer will have. Indifference curves further from the origin give a higher level of utility to the consumer, because higher indifference curves contain more goods. Since more is better, utility is maximized by choosing the combination of goods that places the consumer on the highest indifference curve (the one most distant from the origin) attainable. However, consumers have to operate with a certain income, or under a *budget constraint*. The initial discussion of neoclassical microeconomic theory presented in Chapters 2, 3, and 4 presents the economy as an exchange model without referring to prices. Budget constraints will be added in Chapter 5, when the discussion moves from a pure exchange economy to a market economy, where preferences are indicated by relative prices.

The second assumption, that the consumer acts consistently and rationally, implies that if one combination of beef and Brazil nuts, say commodity bundle 3 in Figure 2.2, is preferred to another combination of beef and Brazil nuts, say commodity bundle 1, and a third commodity bundle, say 2, is preferred to 3, then it follows also that combination 2 is preferred to 1. This relationship is called *transitivity*. The assumptions of transitivity and non-satiation imply that indifference curves cannot intersect. As shown in Figure 2.3, intersecting indifference curves violate the assumption of consistency in consumer choice. In Figure 2.3, both combinations of beef and Brazil nuts given by points 1 and 2 lie on indifference curve I. But point 2 also lies on indifference curve I' indicating the same level of satisfaction as the combination given by point 3 (also on indifference curve '). Since point 2 is common to both indifference curves, this implies that the consumer is indifferent between combinations of goods given by points 1 and 3 (consistency). Yet the commodity bundle represented by point 3 contains more of both beef and Brazil nuts than point 1. Thus the assumption of non-satiation, or more is better, is violated.

The assumption that consumers are in principle willing to trade any good for any other implies that indifference curves have a negative slope. This means that there is some amount of Brazil nuts (or good Y) that would induce the consumer to give up an amount of good beef (good X) to obtain it and vice versa. Therefore, the level of utility a consumer can achieve is determined not only by the absolute amount of goods one has, but also by their relative amounts, (that is, how much of one good there is in comparison to the other). For example, at point 1 in Figure 2.4, with a relatively large amount of beef (good X), this consumer is willing to give up two units of beef in order to get one additional unit of Brazil nuts (point 2), keeping the level of total satisfaction the same. At point 3, with a relatively small amount of beef, the consumer must receive 4 units of Brazil nuts to give up one unit of beef. Obviously, this consumer likes hamburgers and is not willing to give up his beef altogether. As we move from point 1 to point 3, additional units of beef are increasingly more "expensive" in terms of the amount of Brazil nuts this consumer must be offered to agree to the trade. The rate at which a consumer is willing to exchange Y for X is called the *marginal rate of substitution* (MRS). "Marginal" means a small change in something. So the MRS of Y for X is the change in X (in this case, less beef) resulting from a small change in Y (more Brazil nuts) that would keep the consumer's level of satisfaction the same. The value of the MRS is given by the change in X divided by the change in Y (staying on the same indifference curve).

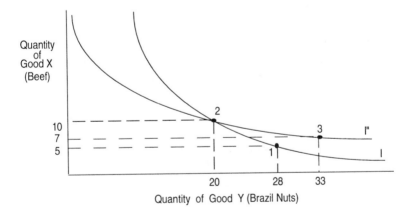

Figure 2.3 Indifference Curves Cannot Cross.

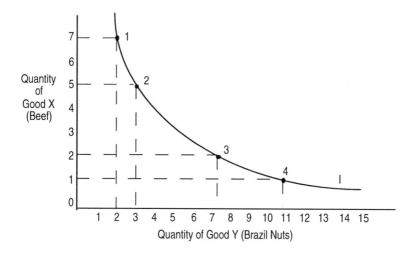

Figure 2.4 The Marginal Rate of Substitution.

Looking at the movement from 1 to 2, the marginal rate of substitution would be $(-2)/1 = -2$, thus indicating that this consumer is willing to give up two units of beef if he is offered one additional unit of Brazil nuts so as to keep his overall level of satisfaction the same. The movement from 3 to 4 shows a marginal rate of substitution of Y for X of $(-1)/4 = -1/4$. Given these amounts of goods X and Y, the marginal rate of substitution shows the number of units of beef $(1/4)$ that this consumer is willing to give up per unit of Brazil nuts received so that overall the level of satisfaction is constant. If we use the symbol "Δ" to indicate "a small change," then we can write the marginal rate of substitution as $\Delta X / \Delta Y$. The expression $\Delta X / \Delta Y$, calculated by dividing the "rise" over the "run" (the change in the variable on the vertical axis, divided by the change in the variable on the horizontal axis), is the *slope* of the indifference curve.

The marginal rate of substitution varies along each consumer's indifference curve, and it also varies between consumers. The different levels of utility different consumers receive from the same combination of goods is the basis for trade. The willingness to trade is also related to another concept in consumer theory, *marginal utility*.

> MARGINAL UTILITY **(MU)** IS THE ADDITIONAL
> AMOUNT OF UTILITY, OR THE ADDITION TO TOTAL
> SATISFACTION, A CONSUMER GETS FROM CON-
> SUMING ONE MORE UNIT OF A PARTICULAR GOOD,
> WHILE HOLDING THE AMOUNT OF ALL THE OTHER
> GOODS CONSTANT.

Additional consumption of beef will generate a change in *total* consumer utility of $(MU_X)(\Delta X)$, which is the change in utility per additional unit of good beef, (marginal utility or $MU_X = \Delta U/\Delta X$) times the number of additional units of beef (good X). Likewise, additional consumption of Brazil nuts (good Y) increases total utility by $(MU_Y)(\Delta Y)$. Moving along an indifference curve requires that total utility remains constant, so that the *decrease* in utility from consuming less beef is exactly equal to the *increase* in utility from consuming more Brazil nuts. This implies that $-(MU_X)(\Delta X) = +(MU_Y)(\Delta Y)$ or, rearranging terms, $-\Delta X/\Delta Y = MU_Y/MU_X$. The marginal rate of substitution, which was equal to $\Delta X/\Delta Y$ or the slope of the indifference curve, $MRS_{Y \text{ for } X} = -\Delta X/\Delta Y = MU_Y/MU_X$, is therefore equal to the ratio of marginal utilities of the two goods.

THE EDGEWORTH BOX DIAGRAM

Armed with the concepts of indifference, marginal utility, and the marginal rate of substitution, we are ready to examine some of the consequences of the neoclassical theory of exchange. To do so, the following section presents the essence of this theory by means of an Edgeworth Box diagram, named after the economist and mathematician F.Y. Edgeworth (1845–1926). While we realize that it may not be easy for those unacquainted with the logic and graphical depiction of economists' ways of thinking, the effort of trying to understand the Edgeworth box analysis is well worth it. We believe the Edgeworth box analysis most clearly presents the conceptual framework of neoclassical economics and its limitations.

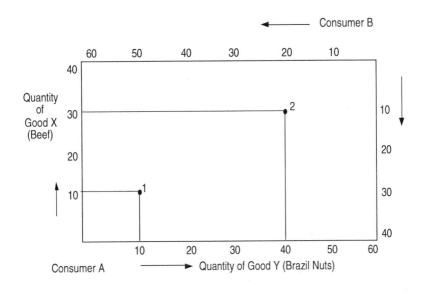

Figure 2.5 Distribution in an Edgeworth Box.

Figure 2.5 shows an Edgeworth box diagram. This diagram is like Figure 2.2 except that it shows two consumers instead of one. The origin in the graph for consumer A is, as before, in the lower left hand corner, but the origin of the graph for consumer B, is in the upper right hand corner. The total amounts of beef (good X) and Brazil nuts (good Y) are assumed as given by the size of the Edgeworth box. In this case, the total amount of beef to be allocated is 40 units, and the total amount of Brazil nuts is 60 units. At point 1 consumer A has 10 units of beef (reading A's amount of X off the left vertical axis) and 10 units of Brazil nuts (reading A's amount of Y off the bottom horizontal axis). Consumer B has the rest of the endowment or 30 units of beef (reading off the right vertical axis from the top down) and 50 units of Brazil nuts (reading off the top horizontal axis). At point 2, consumer A has 30 units of beef and 40 units of Brazil nut, and consumer B has the rest—namely 10 units of beef and 20 units of nuts. So consumer A's utility increases as we move away from the origin up and to the right in the Edgeworth box because this increases the amount of goods going to that consumer.

Consumer B's utility increases as we move down and to the left in the Edgeworth box. Figure 2.6 shows the same graphs adding indifference curves to the diagram. Two indifference curves are shown for each consumer. Of these, consumer B would prefer to be on indifference curve I_b' and consumer A would prefer to be on I_a', or in each case, the one furthest away from the origin.

PARETO OPTIMALITY IN EXCHANGE

Using this exchange model as an analytical tool, it becomes clear how, starting from an initial distribution given by point 1, the trade of beef and Brazil nuts (goods X and Y) can move both consumers to a point in the Edgeworth box where their level of satisfaction is higher. The neoclassical analysis of exchange begins with two very important assumptions.

1. The total amounts of goods X and Y are given as the starting point of the analysis; they are fixed and known to both consumers.

2. The initial distribution of these two goods between the two consumers is also given at the beginning of the analysis.

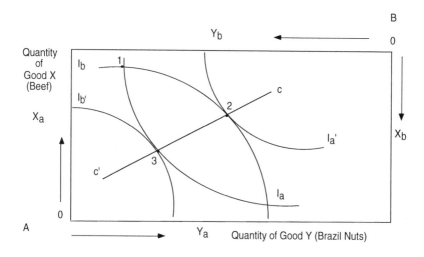

Figure 2.6 Pareto Optimality in Exchange.

Take our example of Carl's beef and Brazil nut coupons. Consider point 1 in Figure 2.6 as showing the initial distribution of these two goods between two of his neighbors, Alex and Bertha. As we move to point 2, the following exchange takes place. Alex has traded some of his beef for some of Bertha's Brazil nuts. Bertha has given up some of her nuts to get more beef. After this trade, compared to the initial distribution at point 1, Alex is better off (on a higher indifference curve), while Bertha is just as well off as before the trade (she is on the same indifference curve). Since the utility of one consumer is increased without reducing the utility of the other, we can say that the initial situation has been improved. If we move from point 1 to point 3, Bertha is better off while Alex is not made worse off. If trading moves the distribution of X and Y to some point on the line CC' between points 2 and 3, both Alex and Bertha will be better off than they were at the initial distribution point 1 (remember that there are an infinite number of indifference curves completely filling the Edgeworth box). Exchange has been motivated by the goal of reaching a higher level of utility or satisfaction than before, indicated by higher indifference curves. The line CC' is called a *contract curve*. Once trading has moved the consumers to the contract curve, Pareto optimality has been achieved.

> **P**ARETO OPTIMALITY IN CONSUMPTION IS ACHIEVED WHEN NO FURTHER TRADING CAN MAKE ONE PERSON BETTER OFF WITHOUT MAKING SOMEONE ELSE WORSE OFF. **A**LL POINTS ON THE CONTRACT CURVE ARE PARETO OPTIMAL.

Along the contract curve, the indifference curves of each consumer are tangent to each other. The contract curve is therefore the locus of all such points where indifference curves just touch. We saw that the MRS is equal to the slope of the indifference curve ($\Delta X / \Delta Y$). At the point where two indifference curves are tangent, the slopes and therefore the marginal rates of substitution between the two goods are the same for both consumers. Thus, along the contract curve the rate at which each consumer is willing to trade one good for another is the same. This gives us the first of three necessary conditions for Pareto optimality.

> ### PARETO CONDITION I: $MRS^A_{YX} = MRS^B_{YX}$
>
> PARETO OPTIMALITY IN CONSUMPTION OCCURS WHEN THE RATE AT WHICH CONSUMERS ARE WILLING TO SUBSTITUTE ONE GOOD FOR ANOTHER IS THE SAME.

If these rates were not the same, further trading could improve the situation of at least one consumer without hurting the other. Notice, however, that each different initial distribution of the two goods may yield a different point on the contract curve, and thus a different point of optimality in exchange. The neoclassical notion of efficiency, as indicated by Pareto optimality, says nothing about the relative desirability of points along the contract curve. A movement from one point to another on the contract curve will make one person better off and make the other person worse off. At point 2 in Figure 2.6, consumer A has relatively more of the two goods, and at point 3, consumer B has relatively more, but we cannot use the Pareto criterion to choose between these two distributions. This illustrates the meaning of "value neutrality" in neoclassical theory. Questions of "fairness" are not addressed by the Pareto criterion. Neoclassical theory states that things can be improved by moving from a distribution off the contract curve to one on the curve.

Pareto optimality is the goal of neoclassical policy. If people are free to trade and are fully informed of the characteristics of the goods available, the end result of each person pursuing his or her own self interest will be a Pareto optimal situation. Consider again our example of Carl's coupons for beef and Brazil nuts. Since he distributed them randomly and unevenly, one person may have 10 of one kind of coupon and 15 of the other, while another person may have coupons for 5 pounds of beef and 20 pounds of Brazil nuts. Almost certainly Carl's friends and neighbors can trade with each other and end up happier than before. People have different tastes and so trades between individuals can give each person more utility. In this example with a small group of people, there are no barriers to trade, there is precise information as to what each person has, and there are no costs involved in making the trades (no *transactions costs*). If trade is allowed to proceed, the end result is a situation of Pareto optimality. No further trading can make one person better off without making another person worse off. This is the kind of situation neoclassical theory describes, and it is this task of distributing goods according to communicated preferences that the market does best. It doesn't matter where the goods come from; it doesn't matter what the initial distribution of these goods is; it doesn't matter how individual tastes are formed or what these tastes are. Given these initial conditions, unrestricted trade will lead to Pareto optimality.

Although the analysis throughout this book will assume only two goods and two consumers, the model can be extended to any number of players and goods. In mathematical notation the utility function will look like

$$U = f(X_1, X_2, X_3,..., X_n).$$

This reads "utility is a function of the amounts of the goods X_1, X_2, up to the last good X_n." As we saw above, more of any good is preferred to less, meaning that the marginal utility of any good X is positive, $\Delta U / \Delta X > 0$. A positive change (an increase) in X will result in a positive change (an increase) in utility U.

CONSUMER THEORY AND THE BIOPHYSICAL WORLD

Even at this early point in our study we can begin to see the implications of the theory of exchange for the valuation of environmental attributes, and the effect of this analytical framework on environmental policy. Consumer theory implicitly assumes that all items which give a consumer utility may be traded in the way described by the indifference curve model. The neoclassical model of exchange takes consumer preferences as given. It is sometimes called the theory of "consumer choice" indicating that consumer preferences, and the choices based on those preferences are what drives the theory. The goal, then, is to maximize utility regardless of how preferences are formed or what the objects of consumer choice are. A concept of limits or of "enough" is foreign to this framework since non-satiation implies that more is always better. The choice framework of demand theory is particularly problematic, since new demands are constantly created as we choose one good over all others. The choice of one good implies the denial of all the other goods that could have been obtained through trade. Since those wants remain unmet, choice creates its own future demand.

The origin and consequences of consumer choices are assumed as given just as the amounts of goods and services that form the initial endowment (e.g., the pounds of beef and Brazil nuts) to be traded. While these endowments and tastes may change over time, in each time frame the current (immediate present) situation determines the dimensions of the Edgeworth box and the location and shape of the indifference curves. In the remainder of this chapter, we will address some of the consequences of these assumptions.

The Lack of Information about the Natural World

The question of how preferences are formed may not be of great significance for most market goods. It may not matter greatly whether people prefer red shirts or green shirts. Regardless of what drives consumer choice, the market will react to consumer demand to supply products accordingly. In some cases, however, simple preferences have serious side effects. The preference for beef over nuts, throwaway razors over reusable ones, or aerosol sprays over pumps, may have significant consequences for the environment. Beef production may require deforestation, while the production of Brazil nuts may not. Throwaway products accumulate in landfills where effluents affect ground and surface water, while reusable products create less waste. Aerosol sprays may contain ozone depleting gases while pump sprays do not. How consumers make decisions and how their preferences are formed, therefore, is relevant.

The level of information available to consumers about the effects of their decisions may be critical for formulating preferences and for environmental policy. For example, according to many biologists, the current loss of biodiversity is the major environmental crisis we face. It is estimated that we have names for only about 10 percent of existing species. Next to nothing is known even about the ones for which we have names. How can consumers make informed exchange decisions between, for example, beef and Brazil nuts when the consequences of their decisions for biodiversity are unknown and probably unknowable? How can policy makers make informed decisions about production priorities or ecologically sound products if the ecological consequences of their decisions are unclear? Many economically valuable plants were considered to be weeds until uses for them were found. Additional criteria are needed to guide both consumers and policy makers as serious negative feedbacks may result from our consumption decisions which go beyond the simple assumptions of exchange without consequences. Uncertainty demands prudent decisions, not careless ones.

The Assumption of Substitutability

At the root of the problem of how various goods and services are valued by neoclassical theory, is the tenet that *all* choices are made in the narrow realm of market exchange. This may make sense for some goods, such as red shirts or blue shirts, which are relatively homogeneous and substitutable. In neoclassical consumer theory, however, there is nothing unique about the utility derived from *any* good. People get utility from open space, clean air, accessible clean water, fertile soil, and so on; but they also get utility from things that diminish the quality of these environmental goods, such as hamburgers, automobiles, and second homes in wilderness areas. In the neoclassical world, each of these goods can be assigned value by determining its exchange value or MRS with any other good. In the world of pure exchange, even environmental features that sustain both the production of consumer goods and the consumer him/herself are treated just as any other market good. As long as they lead to consumer satisfaction, the consumer's task is to balance all competing wants in such a way as to maximize individual utility. Qualitative differences are reduced to quantitative exchange between goods. All other factors affecting utility remain outside, or "external," to the concept of consumer theory (the concept of "externality" will be explored in greater depth in Chapter 6).

There may be cases where a tradeoff between environmental goods and other market goods is legitimate. An individual landowner, for example, may trade off a scenic view for income by

cutting the timber on his land. With most of the serious environmental problems facing us, however, the assumption that all goods
enter the market and are subject to trade on an equal footing with
all others is problematic. When others, including future generations, are affected by individual choices, the level or choice of
consumption cannot be based on individual preferences only. Health
officials, for example, would not set the permissible level of coliform
bacteria based on individual preferences. Initially, the production
and use of aerosol sprays was based solely on individual preferences. Government agreements to phase out the use of aerosol
propellants were only implemented after the scientific community
realized that these sprays were contributing to the destruction of
the earth's protective ozone layer. Likewise, the decision about
how much beef and Brazil nuts to consume is not merely a matter
of individualistic preferences since it affects forest habitats, hydrological cycles, and atmospheric CO_2 concentrations. There are many
examples of economic goods that were originally allocated based
on individual preferences. However, when found to create far-
reaching harmful effects (for example, DDT and PCBs), they were
regulated according to collective, not individual, needs. There are
limits to substitutability. The ability of the biophysical world to
provide a suitable home for humans cannot simply be placed on an
equal footing with market goods and services. The life support
systems of the planet, such as water, air, and living species involved in breakdown, release, and binding of vital nutrients, are
non-substitutable. They must be present in some minimum quantity to make economic activity possible and in fact, to insure human
survival. Likewise, not all human produced goods and services can
be considered on an equal footing. Different goods have different
impacts on the earth's life support systems. The notion of substituting one good for another (e.g., beef for Brazil nuts) based on individual "utility" ranking or "indifference" alone is not an adequate
measure to account for these differences.

Irreversibility, Threshold Effects, and Interconnectedness

The problem of relying on markets to assign value to environmental goods is aggravated by the fact that many decisions which
affect the productive or assimilative functions of nature are irreversible within a relevant time frame. With most market goods,

changes in supply and demand are reversible. If consumers want more TVs at a later time, they can easily be produced, provided of course that the necessary resources have not been depleted, or that the production itself has not proven to have such negative impacts on consumers' health and well-being that it has to be restricted. Goods like air and water quality, or soil fertility, however, are fundamentally different. If a species is driven to extinction, or if the ozone layer is reduced, or if the temperature of the planet rises due to the greenhouse effect, the system cannot be brought back to its original state even if consumers would want the original condition to be restored. Many ecological effects are either altogether irreversible (as in the case of species extinction), or they are irreversible within a relevant time frame (as in the case of ozone depletion and global warming).

The question of irreversibility is related to another critical implication of consumer theory. Environmental goods or attributes are characterized by interconnectedness—what economists call *complementarity*. Complementarity exists with goods like hamburger beef patties and hamburger buns, roasted Brazil nuts and beer, or tennis balls and tennis rackets; pairs of commodities that are used together. The complementarity among environmental entities, however, is of a different order of magnitude. Any ecological system is composed of hundreds of thousands of organisms whose survival is intimately linked to all the others. The exact nature of these links is largely unknown and almost certainly unknowable. Furthermore, there are critical connections between the biosphere, the atmosphere, climate, and hydrological and even geological conditions. If we disturb one of these elements, we disturb all the others. As the conservationist John Muir noted, everything in the universe is connected to everything else. The economic value assigned to beef or Brazil nuts cannot be viewed as separate from the effects of their production and consumption on the rest of the planet. There is a degree of complexity in consumption that cannot be captured by the simple notion of indifference between individual goods and services.

Making choices between various market goods is something all of us do every day. We cannot make the assumption, however, that these choices are a meaningful expression of the value of our consumer goods of choice to other parts of the ecosystem. For this assumption to be valid would require a level of information regarding the consequences of our decisions that goes far beyond the scope of any individual's preference. This is not to say that we could not get closer to optimal exchange decisions if information regarding such consequences was more precise. However, the limits of relying on individual preferences need to be explicitly recognized, particularly when consumer decisions affect the life-sustaining functions upon which present and future generations depend.

Discounting

Some of the biological laws that constitute the parameters of human activity operate on time scales of hundreds or even hundreds of thousands of years. All human economic activity takes place within the boundaries of these laws of physics, chemistry, geology, and biology. However, because market economies are driven by individual decisions made at one specific point in time, parts of our biophysical world that have value over a long stretch of time may be sacrificed for immediate gain under the laws of market exchange. In our beef and Brazil nut tradeoff, we considered only the immediate problem of trading a given amount of these goods. The long-term consequences of our preference for beef, however, may include the cutting or burning of tropical rainforests for cattle ranches to grow the beef. Brazil nuts, on the other hand, may be a rainforest crop that can be harvested without destroying the entire forest. The adverse effect of destroying the rainforest, in terms of biodiversity loss or the contribution to global warming, may not be apparent within the lifetimes of our two consumers, Alex and Bertha. Their choice between beef and Brazil nuts may be made without considering the long-term adverse consequences for the environment since it will not affect them personally.

The present or "now" focus of consumer theory reflects the fact that people would rather have something today than in the future. In economic theory, all goods that give individuals utility at some future date are subject to discounting—that is, they are worth less and less the further into the future we go. Discounting the future allows economists to use the neoclassical model to determine a present rate of exchange for goods delivered at some future date. If some commodity delivered today yields 10 units of utility (*utils*) and if a consumer has a discount rate of 10 percent per year (meaning that something delivered a year from now is worth 10 percent less than if it is delivered today), then the value of that commodity if delivered one year from today is 9 utils. We can then put this discounted value of 9 utils back into the Edgeworth box framework and proceed as before.

There are many important consequences of this "immediate present" orientation as reflected in a positive discount rate. In terms of the social or biological value of ecosystems, it makes little sense to claim that they are worth less in the future. Should environmental policies be formulated based on the assumption that the value of breathable air, drinkable water, or a stable climate continually and sharply declines as we go further into the future? This might make sense for a person who considers nothing but their own finite lifetime measured in a few decades; however, it makes no sense if one is concerned for the human species whose lifetime may be millions of years. Market decisions, unfortunately, are made by individual humans, not the human species, much less other species.

The neoclassical model of consumer theory is independent of time. The immediate present is the sole reference point for the determination of an optimal distribution of goods. There is no past influence asserting its effect on the present conditions of consumption and exchange. Likewise, present consumption and exchange are assumed to have no influence on the future. In reality, however, the consequences of past decisions affect consumer satisfaction just as present decisions affect the future.

MODELS AND REALITY

The neoclassical model of consumer theory tries, as any good model does, to depict reality as accurately as possible while simplifying it so that the model remains easily usable. The exclusive focus of neoclassical economics is the market system, and in some ways its models give a useful description of the way in which markets work and consumer preferences are reflected in the world of market exchange. Connections and feedbacks, that is, the far-reaching consequences of our consumer decisions, are not accurately reflected in market exchange and, therefore, not in the neoclassical description of market exchange. No consideration is given to the absolute scarcity of natural resources, or their value to the planetary life support system. Because market decisions are made by individuals at a point in time, the market economy and the neoclassical model describing it place a lower value on environmental amenities if they are to be enjoyed at some point in the future.

One basic reason our global environment is under assault is the faulty decision-making process of the private market itself with its faulty representation of the life support systems of the planet. The model of Pareto optimal distribution lays bare the inherent limitations of market exchange to assign values to the natural world.

SUMMARY

The neoclassical model of consumption describes the process of exchange of goods that give utility or satisfaction to consumers. It describes the allocation of a given amount of goods with a given initial distribution of them among individual consumers. With unrestricted trade, the final result of exchange will be Pareto optimality. Once this optimal situation is achieved, no additional trading can make one person better off without making someone else worse off.

The trading decisions leading to optimal allocation are made by individuals at a given point in time. Each individual in the neoclassical model begins the process of trading with a given endowment of goods. Neither feedbacks from one consumption period to the next, nor complementarity between the goods exchanged, is considered. No qualitative distinctions are made between ordinary consumer goods and environmental goods essential to ecosystem stability, or between reversible and irreversible decisions. No account is taken of the quality of the information upon which consumer preferences are formed. It is implicitly assumed that individual decisions are sufficient to determine "optimal" levels of distribution and consumption. The environmental policy recommendations of neoclassical economists flow from this narrow model of market exchange. These policy recommendations are limited to actions that improve the expression of consumer preferences in the market. Such policies are restricted, almost exclusively, to ones which seek to facilitate the flow of information from consumers to the market.

This chapter described the theory of consumption in terms of pure exchange, that is, in a barter economy, and the conditions under which it achieves Pareto optimality. This is the basic model or core of consumer theory, and of neoclassical theory in general. Prices are added to this model as indicators or signals of consumer preferences when the system is too large and complicated for individuals to negotiate directly with each other. Chapter 5 takes a closer look at price theory, when the concept of the demand curve is developed from the basic concepts of consumer theory introduced here.

The next chapter extends the model of pure exchange outlined here to describe the activity of the firm, or the production sector. As we will see, the neoclassical theory of production can also be described as a theory of pure exchange. But in place of individuals exchanging goods to maximize utility, we have firms exchanging productive inputs in order to maximize output.

SUGGESTIONS FOR FURTHER READING

Daly, John and John Cobb. *For the Common Good*. Beacon Press, Boston, 1989.

Frank, R. H. *Passion With Reason*. 2nd Edition. W.W. Norton, New York, 1993.

Kennedy, Gavin. *Mathematics for Innumerate Economists*. Holmes and Meier Publishers, New York, 1982.

Nelson, Julie. "The Study of Choice or the Study of Provisioning? Gender and the Definition of Economics," in *Beyond Economic Man: Feminist Theory and Economics*, edited by Marianne Ferber and Julie Nelson, Univ. of Chicago Press, Chicago, 1993.

Stigler, George. "The Development of Utility Theory," *Journal of Political Economy* 59, parts 1 and 2 (Aug./Oct. 1950), 307–327, 373–396.

Veblen, Thorstein. "The Limitations of Marginal Utility" in *The Philosophy of Economics: An Anthology*, edited by Daniel Hausman, Cambridge Univ. Press, Cambridge, 1984.

3

THE THEORY OF THE
FIRM

INTRODUCTION

The firm is the second organizational unit of economic theory. Firms produce the goods and services households consume. They do this by using productive resources, also called *inputs* or *factors of production*. The process by which the firm organizes production is a mirror image of that by which consumers allocate goods to maximize utility (Figure 3.1). The basic concepts in production theory are the *production function* and the *isoquant*. A production function is an equation, a chart, or a graph showing the relationship between the amount of output of some good and the inputs used to produce that good per period of time. In our example, the production of beef requires land, machines such as tractors, materials such as fencing for the cattle pastures, buildings for equipment and meat processing, and so on. Likewise, the production of Brazil nuts requires land for growing the trees, labor for gathering the nuts, and equipment for processing, roasting, and packaging.

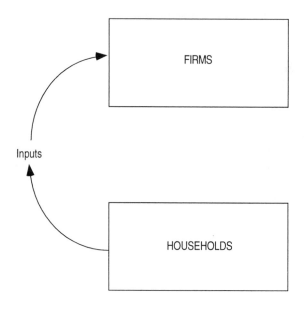

Figure 3.1 The Theory of the Firm in Production.

The simplest and most general production function is written Q = f(K,L). This function indicates that the amount of beef produced (Q) is *a function of* the amounts of capital (K) and labor (L) used in the production process. As in our discussion of utility we will consider here only two goods, beef and Brazil nuts, which are produced in two different firms. Likewise, only two inputs are considered, namely capital and labor, but the analysis can easily be extended to any number of inputs and goods (or firms) by using mathematics instead of two dimensional graphs.

Before we go any further in our analysis we need to clarify the motive behind production theory. Why would anyone raise cattle or grow Brazil nuts? In neoclassical theory, the answer is—*to maximize profits*. The concept of profit maximization will be discussed in more detail in Chapter 5 when prices and production costs are introduced. In the following analysis it is assumed that the goal of the firm is to produce as much output as possible from a given amount of inputs. Just as households seek to maximize utility (satisfaction) by optimally allocating their consumption of various goods and services, firms seek to maximize profits by optimally allocating the different inputs used to produce goods and services. These basic assumptions precede the analysis of Pareto optimality in production:

1. Given an endowment of productive inputs, more output is preferred to less.

2. An increase in inputs leads to an increase in output (in economic jargon, the marginal products of all inputs are positive).

3. Rationality among producers implies that isoquants are downward sloping in the relevant realm of production, that is, that inputs can be substituted.

THE ISOQUANT

> AN ISOQUANT ("SAME QUANTITY") IS A LINE SHOWING THE VARIOUS COMBINATIONS OF TWO INPUTS, FOR EXAMPLE CAPITAL AND LABOR, WHICH CAN BE USED TO PRODUCE A GIVEN QUANTITY OF OUTPUT.

Output levels stay the same as we move along a particular isoquant. Referring to Figure 3.2, as we move up and to the right away from the origin in the graph, output increases from 500 to 1000 tons of beef because there is more of both inputs available to use in production. Isoquants are generally assumed to be downward sloping, which means that capital and labor are substitutable. Labor can be substituted for capital, for example, as more workers and fewer machines are used in a switch to a "low tech" production process. The problem facing the firm is to substitute inputs so as to find the most efficient input combination to produce a certain amount of beef. Not only labor and capital but any particular input has substitutes according to neoclassical theory. It is of course recognized that there are also complementary inputs that must be used together (see appendix). However, the basic concept used in production theory is substitutability.

Figure 3.2 shows an isoquant (I) depicting the combinations of capital and labor needed to produce 500 tons of beef (good X). According to this Figure, a firm producing beef could produce the same 500 tons using either 10 units of capital and 3 units of labor, or using 6 units of capital and 5 units of labor. In contrast to our discussion of the rather vague concept of utility and indifference curves, the isoquant and the underlying production function describing it, say something about physical reality. Figure 3.2 is saying that 500 units of beef can actually, physically, be produced using varying amounts of capital and labor as shown by the isoquant, I.

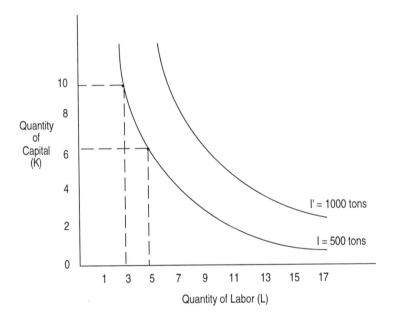

Figure 3.2   An Isoquant.

The ability to substitute one input for another varies along the isoquant. If the firm has a relatively large amount of capital, it can give up a large amount of it for a relatively small amount of labor and keep output at the same level. In other words, eliminating an additional unit of labor becomes more and more "costly" when it comes to the last few units of labor, as, for example, the person pushing the button to start the machine. Shape and location of the isoquant depend on the substitutability of the two inputs and on the technological possibilities available to the firm (for a further discussion of substitutability assumptions underlying the shape of the isoquants, see the appendix at the end of this chapter). If there is a technological improvement in the production of beef such as new feed supplements or new processing equipment, the isoquant shifts toward the origin as shown in Figure 3.3, because fewer inputs are then needed to produce the same amount of beef. After the technological improvement, the same output, here 500 tons of beef, can be produced with fewer inputs of capital and labor. Figure 3.3 shows a parallel shift of the isoquant, but it is recognized that it may become flatter, indicating a capital-saving technological change, or become more steeply sloped indicating a labor-saving technological change.

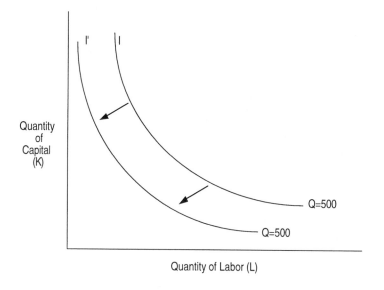

Figure 3.3 A Technological Improvement.

The rate at which one input may be substituted for another, without changing the level of production, is called the *marginal rate of technical substitution* (MRTS). It is given by the slope of the isoquant and is written $MRTS_{L \text{ for } K}$. It is calculated by dividing the "rise" by the "run," or the change in capital divided by the change in labor. Since the word "marginal" means "a small change in," the MRTS indicates the change in capital necessary to offset a small change in labor in order to keep output at the same level; or it might show the additional labor hours necessary to offset a small (marginal) decrease in the number of machine hours used to produce a specific quantity of output; that is, MRTS = $\Delta K/\Delta L$.

A small increase in labor will lead to an increase in output. This is called the *marginal product* of labor, and it can be written as MP_L = $\Delta Q/\Delta L$. Likewise, an increase in capital also leads to an increase in output, or $MP_K = \Delta Q/\Delta K$.

> THE MARGINAL PRODUCT OF AN INPUT IS THE AMOUNT BY WHICH OUTPUT INCREASES AS ONE ADDITIONAL UNIT OF THE INPUT IS USED, KEEPING THE AMOUNTS OF ALL OTHER INPUTS THE SAME.

If output is kept the same (as we move along an isoquant), a firm can offset a decrease in the marginal productivity of labor by an increase in the marginal productivity of capital. We can then write: $MRTS_{LK} = \Delta K/\Delta L = (\Delta Q/\Delta L)/(\Delta Q/\Delta K) = MP_L/MP_K$. This means that the slope of the isoquant ($\Delta K/\Delta L$) is equal to the marginal rate of technical substitution and is also equal to the ratio of the marginal productivities of the inputs. The rate at which one input may be substituted for another in production depends on the relative ability of an additional unit of input to add to total output.

THE EDGEWORTH BOX DIAGRAM FOR PRODUCTION

With this knowledge of production theory, we can now show that unrestricted trade of inputs among firms leads to the greatest total output of goods, given some initial distribution of a given amount of productive inputs. An Edgeworth box diagram similar to the one we used to examine the exchange of goods between consumers shows this. Figure 3.4 shows an Edgeworth box with point 1 indicating an initial allocation of two inputs (capital and labor) between two firms: Firm X produces beef (good X); firm Y produces Brazil nuts (good Y). Both use labor and capital to produce their respective products.

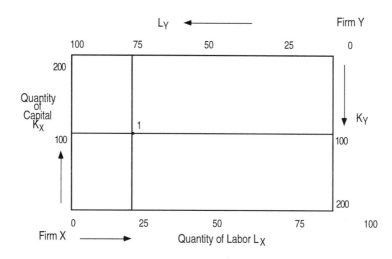

Figure 3.4 An Edgeworth Box Diagram for Production.

Every point in the Edgeworth box diagram depicts some distribution of these two inputs between the two firms X and Y. Figure 3.4 shows that this simple two-good, two-input economy is endowed with 200 units of capital and 100 units of labor. At point 1 each firm has 100 units of capital, firm X has 25 units of labor, and firm Y has 75 units of labor. As in consumer theory, two assumptions precede the analysis of Pareto optimality in production.

1. The initial endowment of capital and labor is fixed at some predetermined amount.

2. The initial distribution of these inputs between the two firms is also given.

Figure 3.5 shows how Pareto optimality in production is reached. The Edgeworth box shows 3 isoquants for firm X and 3 for firm Y. These isoquants show the combinations of capital and labor used to produce beef (X) and Brazil nuts (Y). As one moves away from the origin for firm X, in the lower left hand corner, the output level of beef increases. As one moves away from the origin for firm Y, in the upper right hand corner of the box, the output of Brazil nuts increases. From here we proceed with our analysis in exactly the same way as in the case of consumers trading goods to increase utility.

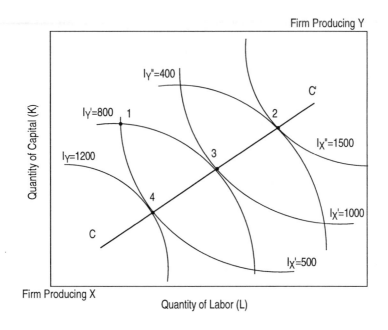

Figure 3.5 Pareto Optimality in Production.

Consider an initial allocation of capital and labor between the two firms as given by point 1. By trading inputs with each other, both firms can increase the total output of this economy. If we move from point 1 to point 3, the production of Brazil nuts (good Y) remains the same at 800 tons (we are still on isoquant I_Y'), while the production of beef (good X) increases from 500 to 1000 tons (as we move from isoquant I_X to isoquant I_X'). If we move from point 1 to point 4 (firm X trades capital for labor, firm Y trades labor for capital), we increase the production of Brazil nuts while keeping the production of beef the same. Moving from point 1 to any point on the contract curve between points 3 and 4 will increase the total output of this economy. Since the Edgeworth box can be pictured as is, completely filled with isoquants, we can construct a contract curve for production showing all Pareto optimal distributions of inputs between the two firms producing beef and Brazil nuts (e.g., at points 2, 3, and 4). Any distribution of labor and capital to the left or right the contract curve CC' is less than optimal, that is, production of at least one firm can be increased by trading inputs. At any point off the contract curve, inputs are used inefficiently.

> **PARETO OPTIMALITY IN PRODUCTION INDICATES THAT NO FURTHER EXCHANGE OF INPUTS BETWEEN FIRMS CAN INCREASE THE OUTPUT OF ONE GOOD WITHOUT DECREASING THE OUTPUT OF ANOTHER GOOD.**

Notice that when we are on the contract curve, the isoquants of the two firms are tangent—they are just touching each other at one point. At these points the slopes of the isoquants (which, as we saw, are equal to the MRTS between inputs), are the same for both products, which gives us the second condition for Pareto optimality.

If resources (inputs) in the economy are free to move from one production process to another, if producers are fully informed about these productive resources, and if producers are able to trade freely, then output will be maximized given an initial endowment and distribution of these inputs.

> **PARETO CONDITION II: $MRTS^X_{LK} = MRTS^Y_{LK}$**
>
> **PARETO OPTIMALITY IN THE EXCHANGE OF INPUTS AMONG FIRMS WILL OCCUR WHEN THE MARGINAL RATE OF TECHNICAL SUBSTITUTION OF THESE INPUTS ARE THE SAME FOR ALL GOODS PRODUCED.**

As in the previous discussion of consumption, nothing has been said about prices. In the neoclassical world prices are merely a reflection of the relative productive characteristics of inputs to be used in production. When a real world situation occurs that is not Pareto optimal, the first instinct of neoclassical economists is to look for some price distortion; that is, some distortion in the way prices reflect the relative value of productive inputs. Yet, even without prices we can already identify the first limitation with the process of exchange itself; neoclassical theory says nothing about the relative desirability of points on the contract curve, that is, nothing about the desirability of more beef and fewer nuts or more nuts and less beef. It simply states that once we are anywhere on the contract curve, the production of one good (beef) cannot be increased any further without decreasing the production of another (Brazil nuts).

THE PRODUCTION POSSIBILITIES FRONTIER

The contract curve for production in Figure 3.5 shows all the Pareto optimal points for the production of beef and Brazil nuts, given a certain amount of inputs and given some initial distribution of these inputs between the two firms producing these goods. We can now take the information given by this contract curve and construct a *production possibilities frontier*.

THE PRODUCTION POSSIBILITIES FRONTIER SHOWS ALL THE EFFICIENT (THAT IS, PARETO OPTIMAL) PRODUCTION POSSIBILITIES FOR THE TWO GOODS, X AND Y. IT SHOWS THE MAXIMUM AMOUNT OF ONE GOOD THAT CAN BE PRODUCED, GIVEN SOME LEVEL OF OUTPUT OF THE OTHER GOOD AND A GIVEN LEVEL OF INPUTS.

Figure 3.6 shows all the most efficient combinations of the production of goods X and Y, given the amount of capital and labor available and given a certain technological capability. Refer back to Figure 3.5 and notice that as we move up and to the right along the contract curve, the production of beef by firm X increases and the production of Brazil nuts by firm Y decreases. This is because more of society's capital and labor resources are now allocated to the production of beef. At point 4 relatively more resources are used to produce Brazil nuts (good Y) than beef (good X); at point 2 relatively more resources are allocated to the production of beef. We can transfer this information to a diagram showing "output space" (the output of beef and Brazil nuts) instead of "input space" (the allocation of the inputs of capital and labor).

In Figure 3.6 points 2, 3, and 4 correspond to the same points on the contract curve in Figure 3.5. All the points on the contract curve

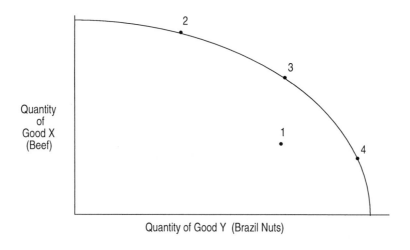

Figure 3.6 The Production Possibilities Frontier.

are Pareto optimal, so, too, are all the points on the production possibilities frontier. Each point shows the maximum amount of one good that could be produced, given the amount produced of the other good and given the endowment of capital and labor inputs. In other words, it shows all the possible combinations of goods X and Y that could be produced if the economy used its resources of capital and labor in the most efficient manner possible. The production possibilities frontier is the key to linking the conditions for Pareto optimality in production and consumption (see Chapter 4).

If the economy is producing a combination of beef and Brazil nuts indicated by point 1 that is below the production possibilities frontier, either (1) there is an inefficient allocation of productive inputs, or (2) there is an "underemployment" of inputs, that is, not all inputs are being used. This second possibility is considered by economists to be wasteful. According to neoclassical economics, it should, therefore, be the goal of society to produce on the production possibilities frontier and thus to use all available resources to produce economic goods. Not to do so represents a loss of production and therefore a loss of utility and social welfare.

PRODUCTION THEORY AND THE BIOPHYSICAL WORLD

The neoclassical model of production described above really has little to do with production. It does not focus on the physical process of transforming inputs into outputs, but instead focuses on the process of allocating given quantities of available inputs among alternative production possibilities. Given a stock of productive resources, some distribution of these resources, and a given technology, the theory shows that through unrestricted trade, resources will be used in the most efficient manner possible so that total output is maximized. The assumption is that more is better and more total output is preferred to less, regardless of the kind of output. Since tradable goods and services are what consumers prefer (the assumption of non-satiation), resources are assigned value only when they enter consumer preferences either directly (as goods) or indirectly (as inputs). The use of resources is not prioritized in any way. All output is on equal footing. In the example of beef and Brazil nuts, capital, labor, and land are used for the production of both goods, and both, therefore, compete for the use of these inputs. The fact that harvesting of Brazil nuts does not require the destruction of tropical forests, while the production of beef does, makes no difference in the valuation of the inputs capital and labor. Both production alternatives, little beef and lots of Brazil nuts or few nuts and lots of beef, assign the same relative value to labor and capital.

In addition, resources only have value if they generate economic benefit. Since more goods means a higher level of utility, the goal of efficiency in production implies that it is desirable to increase output. For some abundant resources, this view of "optimal resource use" may not be a problem. However, when resources are limited, there is a conflict between value assigned in productive use and value resulting from preservation. It is not the forest land itself but the land used for the production of nuts or beef that is useful. It is not labor itself but the labor employed for the production of nuts or beef that is useful.

Resource Scarcity

Although economics is defined as the study of the allocation of scarce resources among alternative uses, the kind of scarcity economists are concerned with is *relative* scarcity, not *absolute* scarcity. The endowment of resources that starts the allocation process between firms constitutes the fixed framework within which allocation takes place. If the amount of capital in the Edgeworth box in Figure 3.5 were reduced to half or even to one one-hundredth of its original amount, the process of exchange of inputs between firms would still proceed until capital is allocated optimally and a situation of Pareto optimality is reached. The Edgeworth box analysis of production shows that the core of neoclassical theory says nothing about absolute scarcity, or the fact that natural resources are finite.

In addition to using inputs to produce economic goods, the production process itself also generates by-products in the form of waste and emissions. Different production processes pose different burdens to the biological world receiving these emissions. This fact also remains outside the focus of the Edgeworth box. As economist Herman Daly points out, there is no notion of optimal scale in relation to the total available amounts of resources. Neither is there any consideration of the waste created during the production process or as a result of using the goods produced. The assimilative capacity of the environment does not enter the analysis either in the calculation of optimal input allocation or in the consideration of the effects of the production process itself (technology). There is no "existence theorem" in this theory that indicates if a given amount of economic activity is compatible with a sustainable environment or with sustainable resource use. The fact that more resource use now results in fewer resources in the future, more human intrusion on other species, more air and water pollution, and even more negative effects on human health, is irrelevant. As the production possibilities frontier shows, efficient resource use means the total use of the endowed resources in any given time frame.

There are, of course, good economic arguments stating that the economy will adjust to scarcities of particular resources as their prices rise. In some cases, for example, with the use of a scarce mineral like copper, increasing prices will call forth substitutes and encourage conservation. Our point is that the substitution effects these arguments refer to take place outside the basic theoretical framework of Pareto optimality just as the side effects of substitution remain "external."

In the Edgeworth box formulation, at each given point in time a new allocation framework is defined that is unaffected by the past and has no direct impact on the future. As in consumer theory, present use alternatives, present information, present technology, and present allocation decisions are the only reference points. This pure time preference supports an analytical framework in which the question of the absolute level of resource availability is not addressed.

The Assimilative Capacity of the Ambient Environment

Resources that are not used for production purposes do not enter the allocation framework. The fact that a production process or the use and extraction of resources affects other non-inputs (such as an economically "useless" species or the assimilative capacity of air or water) also has no effect on the allocation decision. The fact that neither biodiversity nor soil fertility can be maintained if labor and capital are used to produce beef, while the production of Brazil nuts has far fewer negative effects, makes no difference. There are no feedbacks and only a limited notion of complementarity between inputs in the neoclassical framework. Even when vital ecosystem functions are irreversibly damaged, the market can still allocate resources in a Pareto optimal manner. The focus is on analyzing an optimal exchange process, and not on "real" effects. To achieve optimality in neoclassical production theory, it is irrelevant where resources come from, how their use affects noneconomic entities, or where they are discarded after their use.

Discounting Once Again

The discounting problem discussed in Chapter 2 is also present here. According to standard economic analysis, resources delivered at some point in the future are worth less than if they are available now. As before, the process of discounting allows economists to evaluate resources delivered at some point in the future using the same Edgeworth box framework described above. Consider a natural resource such as oil. If a barrel of oil is worth $20 today, with a 10 percent discount rate, it is worth $18 if delivered in one year. We say that $18 is the *present discounted value* of a barrel of oil to be delivered in one year. We could also state this concept in physical terms; 20 barrels of oil today are equivalent to 18 barrels delivered in one year. In the Edgeworth box framework, the future (one year hence) is brought into the present by substituting the number 18 for the number 20, and the analysis can proceed as usual.

As in the case of consumer goods there is a disincentive to conserve productive inputs for the future. The further one goes into the future, the lower the present value of a resource becomes. In our beef/Brazil nut example, *economic* calculations of future damage resulting from the destruction of the rainforest must be discounted. Thus it is possible that the present value of growing beef may be higher than the discounted value of the environmental damage caused by its production. The process of discounting, necessary for economic calculations of costs and benefits, is antithetical to notions of conservation. Things are worth more now than in the future, so there is a disincentive to conserve.

SUMMARY

The neoclassical model of production is a theory of pure exchange. It describes the allocation of a fixed amount of productive inputs among firms with a given initial distribution of these inputs. With unrestricted trade among firms, the end situation will be Pareto optimality in production. When Pareto optimality is reached, no further trading of inputs can increase the output of one good without decreasing the output of another good.

Inputs delivered at some point in the future are discounted, that is, they are worth less and less the further into the future they are received. A key concept underlying the neoclassical theory of input allocation is substitution. A downward sloping isoquant implies that one input can be substituted for another. The slope of the isoquant shows the rate at which one input may be substituted for another and is called the marginal rate of technical substitution. The neoclassical framework of exchange views production as a static, as opposed to dynamic, equilibrium process of allocating a given amount of inputs.

In Chapter 4 we will conclude our discussion of a pure exchange economy by establishing the conditions for Pareto optimality for the entire economy of consumers and producers. This branch of economics is called *general equilibrium theory* or *welfare economics*. In Chapter 5 we will see how a perfectly operating price system is assumed to ensure Pareto optimality.

APPENDIX—A DIGRESSION ON FUNCTIONAL FORM

The mathematical representation of the production function is called the *functional form*. In the discussion of production we used a very general form, merely stating that output per time period was some "function of" the amounts of inputs used, that is, $Q = f(K,L)$. When economists actually estimate production functions to calculate marginal productivity and the marginal rates of technical substitution between inputs, they must specify some mathematical relationship between inputs, and between input and output. These more precise specifications build in assumptions about the degree of substitutability between inputs, returns to scale (the percentage increase in output when all inputs are increased by the same proportion), and other important characteristics of the production process.

Consider a production function of the form $Q = aK + bL$. This is called a *linear* production function, and it would generate isoquants like the ones in Figure 3.7. In this case the inputs of capital and labor are infinitely substitutable. This means that it is possible to produce this good using only labor or only capital. The marginal rate of technical substitution of labor for capital (the slope of a straight line, in this case) is some constant. For example, if the MRTS is one, then one unit of labor can be substituted for one unit of capital at any point on the isoquant. The $MRTS_{L \text{ for } K}$ is always the same no matter what the relative proportions of capital and labor are.

At the other extreme is the *fixed proportions* production function, written

$Q = \min(aK, bL)$.

The isoquants for this production function are shown in Figure 3.8. In this case, capital and labor must be used in some specific ratio. Suppose the units of capital K are tractors and the units of labor are people driving tractors. If there are 10 tractors and 10 people, adding only more people or only more tractors will not increase output (acres plowed). The marginal rate of substitution in this case is zero, and the inputs must be used together in some fixed proportion.

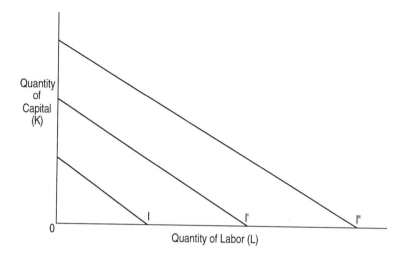

Figure 3.7 Isoquants for a Linear Production Function.

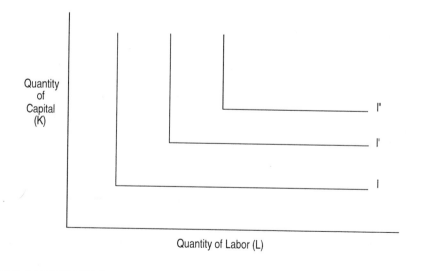

Figure 3.8 Isoquants for a Fixed Proportions Production Function.

Modern production function theory started with the Cobb-Douglas function first proposed in 1924. It was the most widely used production function until quite recently. It is of the form $Q = AK^aL^{1-a}$, where $0 < a < 1$. The isoquant for the Cobb-Douglas function is a rectangular hyperbola as shown in Figure 3.9.

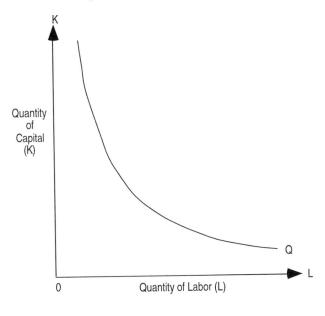

Figure 3.9 Isoquants for a Cobb-Douglas Production Function.

In this case, the ability to substitute one input for another (here labor and capital) decreases as one moves closer to the axes. As discussed in our previous example, to eliminate the last persons picking Brazil nuts or the last persons working the meat processing equipment becomes increasingly "costly" in terms of capital substitution.

The Cobb-Douglas function is an interesting example of the pitfalls of the hidden mathematical properties of a production function. For decades empirical studies were done using this form, which showed that production was characterized by a high degree of substitutability. To measure the substitutability between any two inputs, economists use the "elasticity of substitution." This is a measure of the responsiveness of changes in the relative proportions of inputs used to changes in the relative marginal products of these inputs. For the two inputs capital and labor, the elasticity of substitution (s = σ) is

$$\sigma = \frac{\% \, \Delta \, (K/L)}{\% \, \Delta \, (MPL/MPK)} .$$

For the Cobb-Douglas function, the elasticity of substitution is equal to 1. In the early 1970s econometric studies using the Cobb-Douglas function concluded that non-energy inputs could easily be substituted for energy inputs because the elasticities of substitution between various pairs of inputs were always estimated to be around s = 1. It was not realized then that this conclusion was built into the mathematical structure of the Cobb-Douglas function itself.

In 1961 Kenneth Arrow proposed a more general form of the production function known as the CES, or constant elasticity, function. In this formulation the elasticity of substitution between any two inputs is not constrained to one, as in the Cobb-Douglas case, but it is constrained to be the same between any pair of inputs. For example, if production is a function of capital, labor, and energy inputs, the estimated elasticity of substitution is the same between capital and labor, labor and energy, and capital and energy. Since the CES function a number of more "general" (meaning less restrictive) production functions have been used. The most commonly used form today is the transcendental logarithmic, or translog, function. It places no a priori restrictions on the elasticity of substitution.

The history of the use of production functions in economics can be seen as a steady relaxation of the restrictions on the elasticity of substitution. This has led to a problem in that the more complicated these production functions are, the more sensitive they are to the data used to estimate them. Results obtained using the translog function are notoriously sensitive to even small changes in data.

It is important to keep in mind that isoquants are meant to depict real-world situations, that is, actual technological possibilities. If we use a linear production function in our economic models, we are assuming perfect substitutability between inputs. If we use a fixed proportions production function, we assume that no substitution is possible. In evaluating the ability of the economy to sub-

stitute other inputs for scarce natural resources, we should therefore be aware of the built-in assumptions we make about the
conditions of substitutability that might bias the outcome of empirical studies.

Functional form, therefore, addresses many important issues
about the use of natural resources. Whether or not capital (a *reproducible* input) should be on an equal footing with the *primary* inputs
land and labor is open to question. As Nicholas Georgescu-Roegen
and Herman Daly point out, it is absurd to talk about "substitutability" between capital and natural resources when natural resources are the very basis for producing capital.

SUGGESTIONS FOR FURTHER READING

Ferguson, C.E. *The Neoclassical Theory of Production and Distribution.*
Cambridge Univ. Press, Cambridge, 1975.

Georgescu-Roegen, Nicholas. *Energy and Economic Myths.* Pergamon
Press, New York, 1976.

Heathfield, David and Sören Wibe. *An Introduction to Cost and
Production Functions.* Humanities Press International, Atlantic
Highlands, New Jersey, 1987.

Nicholson, Walter. *Microeconomic Theory: Basic Principles and* Exten-
sions. Dryden Press, Orlando, Florida, 1992.

4
GENERAL EQUILIBRIUM
AND WELFARE
ECONOMICS

INTRODUCTION

In this chapter we bring the producer and consumer side together into one model (see Figure 4.1). This will allow us to establish the third condition for Pareto optimality which explains optimality for the entire economy. The branch of economics concerned with this whole economy perspective is called *welfare economics*. Neoclassical welfare economics is based on three arguments first set forth by Adam Smith in *The Wealth of Nations* published in 1776. These are: (1) humans are motivated by self-interest, (2) if individuals are allowed to pursue their own self-interests, competition will automatically lead to the best situation for society as a whole, and (3) it follows that the best economic policy a government can pursue is to allow the greatest possible freedom for individuals to pursue their own self-interest. The first two of these arguments are the basis for neoclassical theory, and the third argument is the basis of neoclassical economic policy.

Throughout this chapter we continue to deal with a barter or a pure exchange economy. Prices will not be introduced until the next chapter. Among the assumptions made so far are (1) individuals and firms are fully informed of the characteristics of all goods and productive inputs traded, (2) the initial amounts and distribution of these goods and inputs are given, and (3) individuals are able to freely trade their given endowments of goods and inputs. Given these assumptions, free exchange between individual consumers and between firms will lead to a Pareto optimal situation, where no one can be made better off without making someone else worse off, and where resources are allocated so that the production of one good cannot be increased without decreasing the production of another.

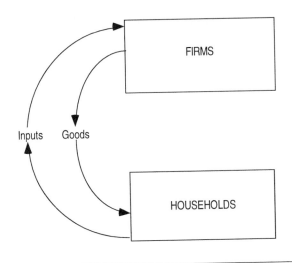

Figure 4.1 The Producer/Consumer Relationship.

GENERAL EQUILIBRIUM IN EXCHANGE

The third condition for Pareto optimality can be derived from the production possibilities frontier described in Chapter 3. We showed that all points on the production possibilities frontier represent different combinations of goods X (beef) and Y (Brazil nuts) that can be produced when society's resources (in our examples, capital and labor) are used in the most efficient manner possible. In other words, it shows all the possible combinations of goods that can be produced when Pareto optimality in production is achieved.

If we randomly pick a point on the production possibilities frontier, we pick a fixed amount of beef and Brazil nuts produced, and thus set the endowment of our society with these two goods. Going back to our assumptions in consumer theory, this means that for each point on the production possibilities frontier, we can construct a particular Edgeworth box diagram showing the particular initial combination of beef and Brazil nuts at that point. If consumers are allowed to trade freely, they will end up somewhere on the contract curve showing all the possible Pareto optimal allocations of the specific amounts of these goods. This procedure is shown in Figure 4.2 which illustrates how we can bring together Pareto optimality in production and in consumption to establish the necessary condition for Pareto optimality in general, that is, for the entire society of consumers and producers.

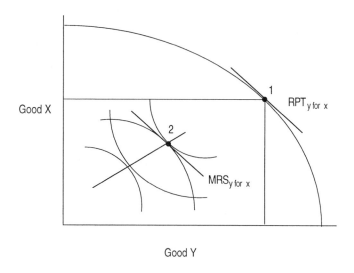

Figure 4.2 Pareto Optimality in General.

The slope of the production possibilities frontier shown in Figure 4.2 gives the *rate of product transformation* (RPT).

THE RATE OF PRODUCT TRANSFORMATION SHOWS THE RATE AT WHICH THE AMOUNT OF ONE GOOD CAN BE REDUCED AND THE AMOUNT OF THE OTHER INCREASED WHILE REMAINING ON THE PRODUCTION POSSIBILITIES FRONTIER. IN OTHER WORDS, IT SHOWS THE RATE AT WHICH ONE GOOD CAN BE GIVEN UP SO THAT MORE OF ANOTHER GOOD CAN BE PRODUCED, CONSIDERING THE GIVEN ENDOWMENTS OF RESOURCES AND THE TECHNOLOGICAL CAPABILITIES OF A SOCIETY.

Recall that the slopes of the indifference curves within the Edgeworth box in Figure 4.2 show the marginal rate of substitution of beef for Brazil nuts, or the relative amount of satisfaction (that is, the ratio of marginal utilities) these two goods generate for our two consumers Alex and Bertha. For the consumer side, we can say that at any point on the contract curve, the marginal rate of substitution between the two goods is the same for both consumers.

The third and final condition for Pareto optimality is:

PARETO CONDITION III: $RPT_{Y\,FOR\,X} = MRS^A_{YX} = MRS^B_{YX}$

PARETO OPTIMALITY FOR CONSUMERS AND PRODUCERS IS ACHIEVED WHEN THE *RATE OF PRODUCT TRANSFORMATION* BETWEEN GOOD X AND GOOD Y IN PRODUCTION IS EQUAL TO THE *MARGINAL RATE OF SUBSTITUTION* BETWEEN THESE GOODS IN CONSUMPTION.

This condition states that resources and goods are optimally allocated when the rate at which beef (good X) must be given up (that is, not produced) in order to free up resources to produce enough Brazil nuts (good Y) to remain on the production possibilities frontier, is exactly equal to the rate at which consumers are willing to substitute beef for Brazil nuts and still maintain the same level of satisfaction (remain on the same indifference curve).

The best way to see that condition III is Pareto optimal is to consider a situation in which it is not met. Assume that the rate of product transformation of beef into Brazil nuts is 1:2, that means if one pound less beef is produced, enough capital and labor is freed up so that producers *are able* to make two more pounds of Brazil nuts. Suppose the marginal rate of substitution is 1:1, that is, consumers *are willing* to give up one pound of beef for one additional pound of Brazil nuts without reducing their overall level of utility, or satisfaction. In this situation optimality has not been reached. A change in production can be made which will increase the total utility of this society. If one pound less beef is produced, enough resources are freed to produce two additional pounds of Brazil nuts, since the RPT is 1:2. But because the consumers' MRS is 1:1, even one extra pound of Brazil nuts in exchange for the one pound of beef given up would keep their overall satisfaction level the same. The one extra pound of Brazil nuts, therefore, increases the consumers' utility level.

Given the assumption that more is always better (non-satiation), this simple two-good (two-firm), two-consumer society is undeniably better off after making these changes, and producing more Brazil nuts and less beef. Only when the rate of product transformation is equal to the marginal rate of substitution is a situation achieved in which no further change can improve the welfare (as measured by efficiency in allocation) of this society. Such a Pareto optimal position is shown in Figure 4.2, where the slope of the production possibilities frontier (the RPT) at point 1 is equal to the common slopes of the indifference curves (the MRS) at point 2 in the Edgeworth box for consumers A and B.

When the third Pareto condition is met, we say the economy is in *general equilibrium*. The word *general* indicates that we are talking about the whole economy of producers and consumers. The word *equilibrium* means that once the third Pareto condition has been established the economy will be stable, unless disturbed by some outside influence, and if disturbed, it will always tend to return to its equilibrium state.

This general Pareto optimal situation is the goal of neoclassical economics. Given some initial endowment of resources (productive inputs) that produce a certain amount of goods and services, and given some initial distribution of the resources among producers, and given an initial distribution of the goods produced among consumers, unhindered exchange will lead the economy to the most efficient allocation of goods and resources.

We can take the notion of Pareto optimality one step further by constructing a *utility possibilities frontier*. A utility possibility frontier is derived in the same manner as the production possibilities frontier we constructed from the contract curve in an Edgeworth box for production. It shows all the Pareto optimal combinations of utility of consumers A and B resulting from various initial distributions of goods X and Y. We can derive such a utility possibilities curve for all the Pareto optimal combinations of goods X and Y that result from the initial endowments of these goods given by all the points on the production possibilities frontier. Figure 4.3 shows several utility possibility frontiers that can be constructed using the information in a contract curve in consumption such as the one shown in Figure 2.6.

We cannot use neoclassical notions of efficiency to say which point on a utility possibilities curve is "best." We can, however, use the information contained in all the utility possibility curves to construct one *grand utility possibility frontier*.

> **THE GRAND UTILITY POSSIBILITIES FRONTIER SHOWS ALL THE PARETO OPTIMAL COMBINATIONS OF UTILITY CONSUMERS MAY DERIVE FROM THE CONSUMPTION OF ALL POSSIBLE COMBINATIONS OF GOODS THAT ARE PRODUCED WHEN INPUTS ARE USED IN THE MOST EFFICIENT MANNER POSSIBLE.**

We can see in Figure 4.3 that if we move from a point such as 3 to point 1, we have increased the utility of Alex while keeping the utility of Bertha the same. Likewise, if we move from point 3 to point 2, we increase the utility of Bertha without hurting Alex. A move to any point between 1 and 2 on the grand utility possibilities frontier, U, makes both Alex and Bertha better off. The grand utility

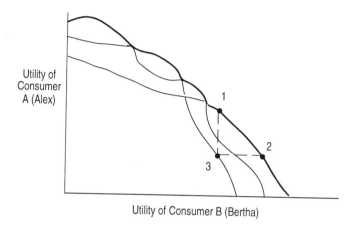

Figure 4.3 The Grand Utility Possibilities Frontier.

possibilities curve is an "envelope" curve constructed by taking the outermost points of the collection of all the possible utility possibility frontiers. Any point interior to the grand utilities possibilities frontier will be less preferred according to the Pareto criterion.

THE SOCIAL WELFARE FUNCTION

We saw in Chapters 2 and 3 that compared to points off the contract curve, any combination of goods or inputs on the contract curve is "optimal." But nothing in the theory allows us to pick the "best" point on the contract curve itself, or the "best" point on the grand production possibilities frontier. This means that using the Pareto criterion alone, we can not make any value judgements about the "fairness" of the distribution of goods between consumers. Likewise, we cannot say anything about whether the resource allocation resulting from individual preferences is really desirable for society as a whole, or whether the level of total production is desirable. To address these questions of scale and distribution, we need to step outside the framework of neoclassical analysis. One analytical tool to address the question of the fairness of the distribution of goods between consumers is to construct a *social welfare function* (W). It may also be called an *iso-welfare* function, since all the points on a given social welfare curve represent the same level

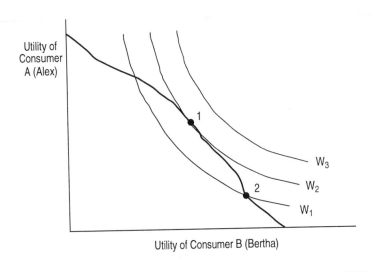

Figure 4.4 The Social Welfare Function.

of total social welfare. Figure 4.4 shows three welfare functions, W_1, W_2, and W_3, which represent something like collective "utility functions" for the entire society.

The social welfare function embodies the welfare judgments of society as to the fairness, or desirability, of the distribution of goods among consumers. Higher utility (resulting from more consumption) is assumed to be better than lower, so the goal is to pick the highest level of welfare possible given the constraint imposed by the grand utility possibilities frontier, that is, by the highest possible level of utility this society is capable of reaching given its production possibilities. In Figure 4.4 this is shown by point 1 (called a *constrained bliss point*), where the social welfare function W_2 is just tangent to the grand utility possibilities frontier. The economy would be better off at some point on the higher social welfare function, W_3, but this function is not attainable given this society's endowment of resources and its level of technology.

The social welfare function allows us to pick the single point on the grand utility possibilities frontier that corresponds to a specific point on the contract curve for consumption that society deems "best." By implication, this also allows us to pick a corresponding point on the production possibilities frontier indicating the most socially desirable mix of goods X and Y. Referring to Figure 4.4, we can see that as we move from one point to another on the utility possibilities frontier, one person is made worse off (her utility is lowered) in order to make someone else better off. For example, by moving from point 2 to point 1, in Figure 4.4, consumer A (Alex) is made better off. But to achieve that higher utility for Alex, Bertha (consumer B) must give up utility and be made worse off. Such a move would not be allowed according to the Pareto criterion. A situation that reduces the utility of one of the consumers in exchange for another's utility increase requires a value judgement about the absolute levels of utility of these consumers that goes beyond the relative utility framework of Pareto optimality.

There is nothing within the basic neoclassical model of producer and consumer behavior that allows us to construct a social welfare function. The neoclassical concept of efficiency can only take us to the production possibilities frontier, or to the grand utility possibilities frontier. To pick out a single point on that frontier from among the infinite possibilities, we must leave the neoclassical framework and include additional considerations about the social desirability of the various possible utility combinations between members of society. The fact that a social welfare function has to be constructed based on an external set of rules, so that a socially optimal allocation of goods and inputs can be determined, is the Achilles heel of neoclassical theory. Once we are forced to come up with rules of choice to pick a particular Pareto optimal combination of goods and a particular distribution of these goods, we can no longer avoid addressing the ethical questions we have dodged so far.

The necessity of a social welfare function was first discussed by the economist Abram Bergson in 1938. Since then a number of economists and social philosophers have suggested rules to construct such a function. Nicholas Kaldor suggested the simple rule that a move from one point to another on the utility possibilities frontier is justified if the person gaining from the move values her gains more than the person who loses values his loss. In other words, if Alex's gain in utility is larger than Bertha's loss, then the redistribution of goods between the two is justified. Tibor Scitovsky amended the Kaldor criterion by adding the condition that after a change is made, we must be sure that society would not be better off by returning to the original situation.

An interesting contribution to the social welfare debate was made by the philosopher John Rawls. Rawls begins with a thought experiment. Suppose that you are to be placed within a society without knowing ahead of time what your social standing would be, that is, what your income would be compared to everyone else? What sort of society would you pick in terms of its income distribution? Would you pick a society with a very unequal income distribution where your chances of being poor are very high, with a small chance of being wealthy, or would you pick a society with a relatively egalitarian income distribution? Rawls argues that most people would pick the latter. He argues further that we should construct our social welfare function on the basis of providing as much income equality as possible, until we reach the point where a move to more equality would reduce the total output of society and reduce the income of the worst-off person. Rawls' argument assumes that people are risk averse and thus unwilling to take the chance of ending up poor, or that they are altruistic and care about another's fate, not just their own.

GENERAL EQUILIBRIUM THEORY AND THE BIOPHYSICAL WORLD

While the welfare considerations that enter the general equilibrium conditions of Pareto Optimality force us to leave a strictly neoclassical world, the assumptions made in that world still enter the general equilibrium analysis unchanged. The relevant time frame is still the immediate present, consumer tastes and production technologies are given, and place as location and social and ecological context is not considered. The analysis is assumed to be universally applicable, and the consequences of changing the starting conditions, such as the initial distribution of goods or inputs, remain outside the framework of optimality. In other words, the broader consequences of individual actions are not considered in the general equilibrium framework. There are no feedbacks in this analytical framework and therefore no need for caution, no need for future orientation, and no need for prevention. The better-off/

worse-off parameters of Pareto optimality are still determined by the quantity of goods available for consumption, resource efficiency is still determined by currently known use alternatives, and the relevant time frame is still the immediate present disconnected from past or future.

The discussion of what determines social welfare has been driven by the goal of material accumulation. More is better while qualitative differences are ignored. The limits to this notion of welfare are becoming increasingly evident. No one would deny that basic material needs must be met in order to achieve some minimal socially acceptable level of welfare. But food, clothing, shelter, or even cars and VCRs are not all that determines our well-being as individuals or as a society. A recent survey conducted in Japan, the country celebrated for its miraculous achievements in economic efficiency and growth, speaks to the problem of adequate welfare measures. In a broad-based survey, the Japanese were asked to choose the two most important social changes from a list of ten. Of the 68 percent who responded, 53 percent thought Japan had transformed itself from a poor into a rich nation, and 46 percent said Japan was no longer thrifty. However, fewer than 3 percent of the respondents thought Japan had become a happier nation.

Our well-being is affected by social structures and support systems like families, neighborhoods, and social context, by the well-being or suffering of others, by the quality of our natural environment affecting our ability to use rivers for swimming, parks for walking, and streets or backyards safely for children's play, *and* by the material things available to us. The costs of social and environmental change may be qualitative, as in lost comfort levels or increased stress, but they may also be quantifiable, as in increased health care or security needs.

What does all this mean for our simple example of Brazil nuts and beef (two goods), and capital and labor (two inputs)? If more people can be fed by producing Brazil nuts than beef, the two cannot be simply equated as generating comparable utility levels since the social welfare effects of the consumptions of these two goods are very different. Likewise, if the same number of calories can be produced from Brazil nuts instead of beef with less land being used or destroyed, smaller losses in biological diversity and less soil erosion and fertility loss, then the use of inputs for beef production results in less welfare, particularly for future generations. If traditional practices of producing beef or Brazil nuts preserve soil and water quality and strengthen social support systems, then the lower yields resulting from such traditional practices may be socially advantageous rather than negative. If an increase in beef production benefits those whose calorie intake is already above 3,500 calories per day, while an increase in the production of Brazil nuts benefits those whose daily calorie intake is much lower, the social evaluation of beef versus Brazil nuts is complicated by the question of who benefits and who loses.

All these examples show the limits of a social welfare function derived from individual tastes as the basis for welfare maximization. They also point to the complex questions involved in interpersonal comparisons of welfare. Can one person's benefits outweigh another person's loss? Despite the fact that neoclassical economists would generally argue that it is impossible to compare interpersonal utility levels, it is recognized that the definition of a social welfare function is necessary as a starting point for general equilibrium analysis.

Social Welfare and Ethics

Humans are social beings. What happens to others affects us. Humans act as social agents embedded in a social context. This has led to a somewhat redefined version of the neoclassical notion of social welfare based on strictly individual preferences. Altruism, for example, is considered one kind of individual preference, as are sadism or masochism. One might also argue, for example, that social inequality, which leads to unrest or a level of air pollution that impacts ones quality of life, would be avoided by a rational-acting, individual interest maximizer.

Some have applied Rawls' understanding of utilitarian ethics to the question of intergenerational environmental equity. If we did not know which generation we would be placed in, how would this affect our attitude toward resources use or pollution? The assumption is that if we thought we might be placed in a society that exists 100 years from now, we would be more likely to be concerned with protecting the biosphere and preserving natural resources. Others, however, suggest that individual preferences are not enough to address the dilemmas and tensions between individual decisions and their impact on social and ecological contexts.

Time, however, does seem to influence our perception of individual versus social interests. A short-term view seems to undermine more altruistic or ecologically conscious behavior. Game theory experiments testing strategic individual behavior confirm the importance of time frame. Experimental results have shown that under conditions of long-term durable relationships, cooperative strategies were far more successful than competitive ones. Transient, short-term relationships, on the other hand, seem to undermine the benefits of reciprocal and mutual solutions.

Cultural differences, too, determine the ways in which individual versus social and ecological benefits are perceived and evaluated. In many societies, Adam Smith's understanding—that the individual pursuit of self-interest also leads to the best interest of society as a whole—would be turned around. The individual's well-being is intricately connected to the well-being of the community, and thus the welfare of the whole is decisive. Examples of a mutual and reciprocal understanding of the individual as part of a larger social and ecological context are found in the belief systems of indigenous peoples. What happens to nature is inseparably connected to the fate of humans. The neoclassical framework asserts not only a particular kind of economic understanding but also a particular cultural perspective of the relationship between individuals and their social and ecological context.

In many ways the global environmental problems we face have added new fuel to the welfare discussion. They have added a new dimension to the interconnectedness between individual and social context. The medium by which we are connected in a very real way over time and over space is the global ecosystem we share. The African woman who has to walk further to get water and work longer hours on drought-affected soils suffers from the consequences of climate changing emissions that stem from U.S. or European factories, heating systems, and automobiles. But the North is also affected by the low-tech inefficient and high-emission, coal-burning plants that generate electricity in many parts of the world where low emission technology is unaffordable. However, adding ecological considerations does not solve the problem of ethical dilemmas raised by the question of social welfare. It merely enlarges the dimensions of the dilemma. All may be affected by such global problems as climate change, ozone depletion, or biodiversity loss. But rather than equalizing them, environmental consequences are likely to exacerbate social inequalities. Examples of toxic waste sites located in minority neighborhoods, waste shipments to so-called third world countries, or the inability to cope with mounting health effects have made it clear that the poor are also more likely to suffer from the consequences of environmental degradation and unsustainable management practices. The question of what defines a socially optimal level of production and consumption is not solved as we confront their effects on the biophysical world, but it may well be brought into sharper focus.

Beyond Human Welfare

Another question the social welfare framework avoids is how we consider the welfare of the non-human, biophysical parts of our world. Do they deserve their own consideration, or do we evaluate them simply based on their usefulness to humans and their impact on human well-being? In the 1930s Aldo Leopold called for a "land ethic" that would respect the rights of nature. This point of view has continued to grow into today's environmental movement with organizations whose members number in the millions. Calls for the ethical treatment of nature are becoming more and more accepted as a result of the growing scientific evidence blurring the distinction between humans and the rest of the animal kingdom.

The ecological ethic of ecofeminism is an ethic of eco-justice, which focuses on the links between social domination and the domination of nature. It sees the roots of the dual oppression of exploited humans and exploited nature in the separation of nature and culture established by the scientific revolution, patriarchal religion, and the dominant psychology of a rights-based rather than a responsibility-based ethic. For the social welfare of humans to be fully considered, the welfare of nature and the links between nature and humans have to be reevaluated.

SUMMARY

The final step in constructing the general equilibrium of neo-classical theory integrates production and consumption into a single framework. In doing so we keep all the assumptions we have made so far—more is better, present tastes and priorities, current knowledge, and technological sophistication determine the better-off or worse-off of neoclassical optimality. In order to address the question of how to determine the most socially desirable Pareto optimal point of production and consumption, our analysis threw us headlong into a discussion of the social welfare function.

The social welfare function has an uneasy place in neoclassical theory. Although it is recognized that such a function is theoretically necessary as a starting point for general equilibrium analysis, neoclassical economists generally argue that interpersonal comparisons of utility cannot be made. In the neoclassical world, the sanctity of selfish, individual decisions dictates that a social welfare function should be based on individual preferences (although "altruism" is recognized as one kind of individual preference). In reality, however, all societies implicitly construct a social welfare function when they make political decisions affecting income distribution and the use of natural resources. Various kinds of transfer payments to the poor, the graduated income tax, environmental regulations, and zoning laws, all embody some notion of a social welfare function. In all these cases, society is taking something from one group and giving it to another, for the good of society as a whole. Apart from the ethical questions involved in determining a society's welfare function, there are serious measurement problems as well. How can we measure a society's welfare and what criteria do we apply? This question can not be answered by economics alone but requires that deep moral and ethical questions be addressed.

The necessity of a social welfare function opens the door to all sorts of environmental policy questions. Is there a "common good" in protecting certain environmental features that should override individual preferences? Should the social welfare function be broadened to include future generations? Should the social welfare function take into account the well-being of other species? Despite its claim to value neutrality, the social welfare function makes explicit that the analytical framework of neoclassical theory has a particular ethical basis; the ethics of a self-interest-oriented individual at a given point in time.

The establishment of the conditions for general equilibrium in a pure exchange (barter) economy completes the first part of the neoclassical analysis of the economy. The next part, covered in Chapter 5, establishes the conditions under which a perfectly competitive economy will ensure that Pareto optimality is met, that is, when a perfectly operating price system will exactly duplicate the result achieved in the barter economy described in this chapter.

SUGGESTIONS FOR FURTHER READING

Booth, Douglas. *Valuing Nature: The Decline and Preservation of Old-Growth Forests.* Rowman & Littlefield, Lanham, Maryland, 1994.

Bormann, Herbert and Stephen Kellert. *Ecology, Economics, Ethics: The Broken Circle.* Yale Univ. Press, New Haven, Connecticut, 1991.

Hirsch, Fred. *Social Limits to Growth.* Harvard Univ. Press, Cambridge, Massachusetts, 1976.

Leopold, Aldo. *A Sand County Almanac.* Oxford Univ. Press, New York, 1966.

Ostrom, Elinor. *Governing the Commons: The Evolution of Institutions for Collective Action.* Cambridge Univ. Press, New York, 1990.

Quirk, James and R. Saposnik. *Introduction to General Equilibrium Theory and Welfare Economics.* McGraw-Hill, New York, 1968.

Rawls, John. *A Theory of Justice.* Harvard Univ. Press, Cambridge, Massachusetts, 1971.

Sagoff, Mark. *The Economy of the Earth.* Cambridge Univ. Press, New York, 1988.

Schor, Juliet. *The Overworked American.* Basic Books, New York, 1991.

Scitovsky, Tibor. *The Joyless Economy.* Oxford Univ. Press, New York, 1976.

Sen, Amartya and Bernard Williams, (eds.). *Utalitarianism and Beyond.* Cambridge Univ. Press, New York, 1982.

Shiva, Vandana. *Staying Alive. Women, Ecology, and Development.* Zed Books, London, 1989.

Society and Nature, Special Issue on "Feminism and Ecology," 2 (1), 1993.

See also the many articles on economics and ethics in the journals *Ecological Economics, Environmental Ethics, Environmental Values, The International Journal of Social Economics,* and *The Review of Social Economy.*

5

INTRODUCING PRICES: PARETO OPTIMALITY AND PERFECT COMPETITION

INTRODUCTION

So far we have described market exchange and the conditions for Pareto optimality without referring to prices. The markets described in previous chapters are the face-to-face exchange markets of a barter society. Now we add monetary flows to our circular flow model of the economy (see Figure 5.1). Through the price system, monetary flows are established that transfer money from households to firms in exchange for goods and services produced. Firms, in turn, transfer money to the households in exchange for productive inputs. The purpose of the price system in neoclassical theory is to duplicate the Pareto optimal conditions of efficiency in a barter economy. Since direct negotiation and exchange between buyers and sellers in a complex society is impossible, a means of indirect exchange is needed. As a result, markets must depend upon a system of relative prices to communicate the preferences of consumers and producers. According to the ideal conditions of neoclassical theory, the market system works like an auctioneer who communicates the values producers and consumers assign to productive inputs and market goods and services. Prices in this system signal the marginal utility consumers gain from the consumption of a good, and the marginal costs of the inputs used to produce these goods. In this chapter, we will show that in an economy with a perfectly operating price system, the three conditions of Pareto optimality will be achieved.

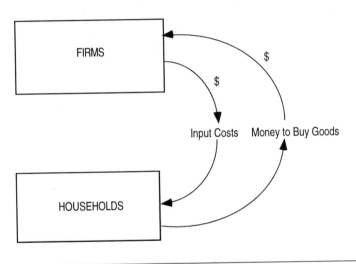

Figure 5.1   Monetary Flows within the Economy.

Now let us go back and see how we arrive at these new concepts of marginal cost and product prices that allow us to establish the Pareto conditions for general equilibrium in a market instead of a barter society.

PRICES IN CONSUMPTION: THE BUDGET CONSTRAINT

Prices are incorporated into the framework of consumer decision-making by means of the consumer's *budget constraint*. If M is the total amount of money a consumer has to spend on beef (good X) and Brazil nuts (good Y), and if the prices of these goods are P_X and P_Y, respectively, then the budget constraint can be written as $M = P_X X + P_Y Y$. Suppose the total amount of money for food available to one of the consumers, say Bertha, is $1000. The price of beef is $10 per pound, and the price of Brazil nuts is $5 per pound. It follows that Bertha can buy a maximum of 100 pounds of beef or 200 pounds of Brazil nuts or some combination of the two. In Figure 5.2, the intercepts of the budget line with the X (beef) and Y (Brazil nuts) axes of the graph show the amount of each good that could be purchased if the entire budget was spent on only one good (M/P_X = 100 pounds and M/P_Y = 200 pounds). Connecting these points gives all the possible combinations of beef and Brazil nuts one could buy with $1000. The slope of the budget line ($\Delta X/\Delta Y$) is equal to the rise over the run or $(M/P_X) \div (M/P_Y)$. Thus we can say that the slope of the budget line is equal to the price ratio of the two goods, or $\Delta X/\Delta Y = P_Y/P_X$.

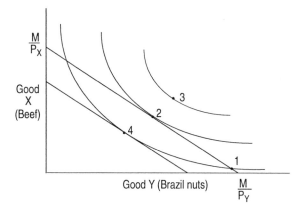

Figure 5.2 Maximizing Utility Subject to a Budget Constraint.

Considering her budget constraint, our consumer Bertha can now only reach a utility level (indifference curve) that is on or below the budget line. Thus the highest possible indifference curve or the best possible combination of beef and Brazil nuts she can choose is point 2 in Figure 5.2. This is where the highest possible level of satisfaction is reached given the budget constraint she faces, or the point of maximum utility under constraint. At point 2 the indifference curve is just tangent to the budget line. Theoretically, she could pick any other combination of beef and Brazil nuts on or below the budget line. For example, the combination shown by point 1 in Figure 5.2 could be purchased. At that point, however, Bertha does not maximize her level of satisfaction. Her utility can be increased (a move to a higher indifference curve) by purchasing the combination of beef and Brazil nuts indicated by point 2. A move to point 3 would give her an even higher level of utility but it exceeds her budget. The higher level of utility represented by point 3 could only be reached if her budget increased. A proportionate increase in the price of both goods, on the other hand, would force Bertha to a lower utility level, since fewer goods can be purchased as prices increase (point 4 in Figure 5.2).

Recall from Chapter 2 that the marginal rate of substitution is equal to the slope of the indifference curve ($\Delta X/\Delta Y$), which is equal to the ratio of the marginal utilities of these two goods ($\Delta U/\Delta Y$) ÷ ($\Delta U/\Delta X$). The slope of the indifference curve shows the rate at which the consumer is willing to give up beef in exchange for an additional unit of Brazil nuts, so that the total utility level is kept the same. Given a budget constraint, the highest possible level of utility a consumer can reach is where the slopes of the budget line and the indifference curve are equal (the tangent point). We can then say that $MRS_{y \text{ for } x} = \Delta X/\Delta Y = P_Y/P_X$. This is the first step in establishing Pareto optimality in a perfectly competitive market economy.

THE DEMAND CURVE

Two things may change the amounts of beef and Brazil nuts (goods X and Y) that maximize a consumer's utility under a budget constraint. These are a change in the relative prices of the goods (P_Y/P_X) and a change in budget or money income (M). This assumes that everything else that might influence utility, such as preferences, accessibility of the goods, information about the goods, etc., stays the same. Economists call this the *ceteris paribus* assumption, which is Latin for "all other things remaining equal or unchanged."

Figure 5.3 shows the effect of a price increase for Brazil nuts from P_Y ($5) to $P_Y{'}$ ($10), if the amount of money in the consumer's budget and the price of beef stay the same. This increase in price will *(ceteris paribus)* decrease the amount of Brazil nuts that can now be purchased. If Bertha spends all her money on Brazil nuts, she can now no longer buy 200 pounds (M/P_Y), but 100 pounds of

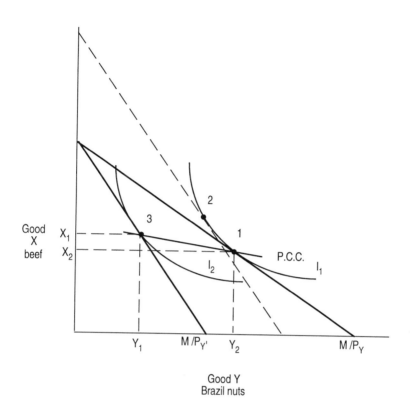

Figure 5.3 The Effects of a Price Increase in Good Y.

nuts (M/P_Y'). An increase in the price of Brazil nuts will therefore cause the budget line to get steeper and shift toward the origin as shown in Figure 5.3. Two things are apparent here. First, Bertha's utility has decreased—she can no longer "afford" indifference curve I_1 but has to move to a lower indifference curve I_2. Since the budget (money) has stayed the same while the price of one of the goods has increased, the total amount of goods that can be purchased decreases. We can say that *real income* (purchasing power) has fallen, and since more is better, utility has decreased. The second thing we see is that the price ratio between beef and Brazil nuts has changed. Brazil nuts are now relatively more expensive. This translates into a more steeply sloped budget line (see Figure 5.3). Both of these things, the decline in real income and the increase in the relative price of Brazil nuts will cause a decline in the amount purchased, from Y_2 to Y_1 in Figure 5.3. These two effects are called the *income effect* and the *substitution effect*. The income effect of a price change means that less of a good is purchased as its price increases because the real income (purchasing power) of the consumer decreases (a move from point 2 to point 3 in Figure 5.3). The substitution effect of a price change means that as the price of Brazil nuts increases, they become relatively more expensive compared to beef, and consumers will switch from Brazil nuts to beef (a move from point 1 to point 2 in Figure 5.3).

The line in Figure 5.3 labeled P.C.C. connects the old and the new best possible (utility maximizing) combinations of beef and Brazil nuts our consumer Bertha can buy given her budget constraints. P.C.C. is called a *price consumption curve*. It shows how the utility maximizing consumption level of good Y (Brazil nuts) varies due to a change in the price of Y. Figure 5.4 uses the information we gained from the price consumption curve to construct the most widely used diagram in economic theory, the *demand curve*. The demand curve shows the relationship between the price of a good and the quantity demanded of that good. It is downward sloping, which indicates that the relationship between price and quantity is negative. The higher the price the less of the good that will be demanded *(ceteris paribus)*. This inverse relationship between price and quantity is called the *law of demand*. The shape of the demand curve gives information as to how the quantity demanded of the good changes in response to changes in its price. Point 1 in Figure 5.4 shows the quantity of Brazil nuts at point 1 in Figure 5.3 that could be purchased at $5.00 per unit. Point 3 in Figure 5.4. shows the utility maximizing quantity of nuts purchased after the price increase (point 3 in Figure 5.3).

The movement along a demand curve for a good such as Brazil nuts is caused by a change in the price of Brazil nuts and is referred to as a *change in quantity demanded*. Changes in the price of related goods or in income cause a shift in the demand curve or a *change in demand*. For example, as the price of Brazil nuts changes, so will the amount of beef that maximizes the consumer's utility (points X_1 and X_2 in Figure 5.3). This is due to the *substitution effect*. If the price

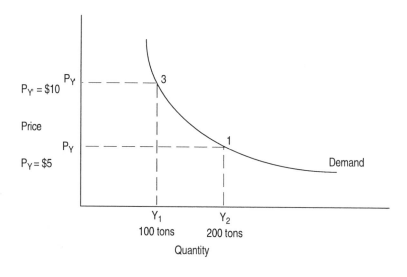

Figure 5.4   The Demand Curve for Brazil Nuts.

of Brazil nuts goes up, more beef is demanded even though its price stays the same. This causes a shift in the demand curve for beef. Such a shift in the demand curve can be caused by changes in the price of a related good, in consumer taste, or in income.

The demand curve is based on the theory of consumer behavior we explored in Chapter 2. It is derived from the notion of indifference and individual utility maximization, to which we added the budget constraint. To get from a demand curve for an individual consumer to the market demand curve for a commodity we simply add up all the individual consumers' demand at each price level.

PRICES IN PRODUCTION: THE COST CONSTRAINT

According to neoclassical theory, the decision about how much to produce and which inputs to use in the production process is made in a similar fashion as consumption decisions are made by consumers, namely, by considering the firm's production budget and input costs. Input costs are depicted by an isocost line (C) as shown in Figure 5.5 showing the various amounts of labor and capital that a firm can obtain by investing its entire production

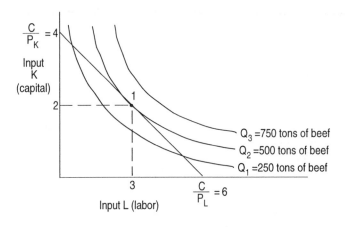

Figure 5.5 Maximizing Output Subject to a Cost Constraint.

budget (C) of say, $60,000 per year. Given input costs of $10,000 per unit of labor per year and capital costs of $15,000 per unit of capital per year, the firm could use a maximum of 6 workers (C/P_L) and 4 units of capital equipment (C/P_K). The slope of the isocost line is $\Delta K/\Delta L = (C/P_K) \div (C/P_L) = P_L/P_K$. The largest possible amount of output that can be produced, given the budget restrictions and input costs, is point 1 on isoquant Q_2 in Figure 5.5. It shows that the firm could produce a maximum of 500 tons of beef. It would do so by hiring 3 workers and buying 2 units of capital equipment. In point 1 (Figure 5.5) the isoquant is just tangent to the isocost line. Recall from Chapter 3 that the Marginal Rate of Technical Substitution (MRTS) is equal to the slope of the isoquant $(\Delta K/\Delta L)$, which is equal to the rate at which labor has to be increased in exchange for a decrease in capital so that the same amount of beef can be produced $(\Delta Q/\Delta L \div \Delta Q/\Delta K)$. Given the production budget restriction, the slope of the highest possible isoquant that can be reached is equal to the slope of the isocost line (tangent point 1 in Figure 5.5). Thus we can say that MRTS $_{L\,for\,K} = \Delta K/\Delta L = P_L/P_K$. This is the second step toward establishing Pareto optimality considering input prices. It holds for all goods produced.

The optimization point we identified in this condition of Pareto optimality should not be confused with profit maximization. Profit maximization means that we maximize the difference between total revenue and total cost. This is not the same as producing the highest possible level of output.

THE SUPPLY CURVE

Since neoclassical theory assumes that producers seek to maximize profits, just as consumers seek to maximize utility, two new concepts need to be introduced here: the cost and revenue functions. Profits are the difference between revenues and costs of production. In our beef example, total revenue is the amount of beef produced and sold (X) times the price of beef (P_X) or TR = P_XX. Economic theory distinguishes between two kinds of production costs—fixed and variable costs. The general assumption behind those two cost concepts is that the firm has less flexibility in production decisions in the short run than in the long run. Many factors such as buildings (i.e., a meat processing and packaging plant), the land on which the production site is located (i.e., the cattle's grazing land) or equipment (i.e., tractors, packaging machines) cannot be easily changed; they are "fixed" in the short run. The "overhead" costs associated with such items are relatively unaffected by the amount of beef produced. Other inputs, such as the number of workers employed, the amount of fertilizer or diesel fuel used or the feeding supplements given to the cattle, can be altered much more easily. They are considered variable inputs that change with the amount of beef produced. The more cattle production increases, the more feed is required, the more fence repairs are required, and the more workers are needed. The total cost of beef production (TC) is, therefore, made up of the fixed costs (F) that are independent of the amount of beef produced, plus the variable costs per unit of output (V) times the amount of beef produced, or TC = F + VX.

Since we assume that production takes place under the ideal conditions of perfect competition (which we will define below), product prices (P_X) are set by the interaction between consumers and producers in the market. Likewise, individual producers buy such a small portion of inputs that they cannot affect input prices. In our total cost and total revenue equations, prices cannot be altered by the producer. Consequently, production decisions are simply a matter of deciding how much to produce, given input and product prices.

Just as the production function $Q = f(K,L)$ shows output as a function of inputs, the cost function shows production costs as a function of output $TC = f(Q)$. Figure 5.6 shows the general shape of a total cost function, TC. The total cost function rises at first, levels off and finally increases more rapidly. The shape of the total cost curve is due to changes in efficiency at various output levels. The total revenue curve is a straight line because the product price is assumed to be the same regardless of how much is sold by this producer (assuming perfect competition). The profit maximizing output level is where the distance between TR and TC is the greatest. Figure 5.7 shows revenue and costs per unit of output produced. It shows that at a product price of P_1, the profit maximizing output level is reached when 500 tons of beef are produced. For any output level above 500 tons, the additional revenue generated per unit of output (*marginal revenue*, MR = $(\Delta TR/\Delta Q)$) is less than the additional production costs caused by the production of an additional unit of output (*marginal cost*, MC = $(\Delta TC/\Delta Q)$). Increasing production beyond this point would not make economic sense.

The fact that MR has to be equal to MC is not sufficient to determine supply. Only if revenues are high enough to cover the

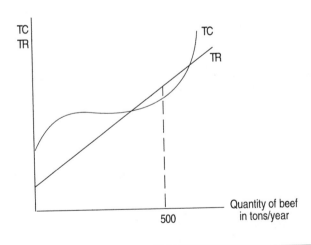

Figure 5.6 Total Revenue and Total Cost Curves.

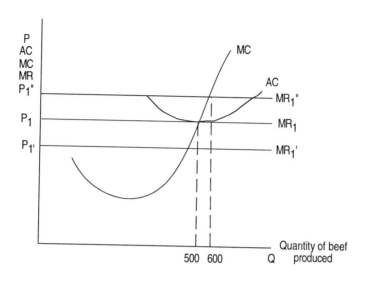

Figure 5.7   Marginal Revenue, Marginal Cost and Average Cost of Production.

per unit costs of production will additional output be produced. If the cost to produce one additional pound of beef (MC) is $4.00 and the additional revenue (MR) is $4.00 per pound ($P_1$ in Figure 5.7), 500 tons will be produced. If beef prices increase to $5.00 per pound ($P_1''$ in Figure 5.7) the profit maximizing level of output increases to 600 tons. If the price for beef, however, drops to $3.00 per pound ($P_1'$ in Figure 5.7), while the average costs of producing one unit of beef is $4.00, production will stop.

> **MARGINAL REVENUE IS THE ADDITION TO TOTAL REVENUE RESULTING FROM THE PRODUCTION OF ONE ADDITIONAL UNIT OF OUTPUT.**
>
> **MARGINAL COST IS THE ADDITION TO TOTAL COST INCURRED FROM THE PRODUCTION OF ONE MORE UNIT OF OUTPUT.**

While it may seem that at price level P_1 (Figure 5.7) this producer realizes no profits at all, this is not so. Economists distinguish between *accounting profit*, what most of us think of when we hear the word profit, and *economic profit*. Economic profits are the profits generated over and above those obtainable in the best alternative in the economy. They include entrepreneurs' earnings and returns on their capital investments. The assumption here is that economic profits of zero exist when there is no better alternative in the economy for the entrepreneur to invest personal labor or capital, that is, economic profits include opportunity costs. If a particular firm is earning an 8 percent rate of accounting profit and the average rate of accounting profit in the economy is 8 percent, then the firm's economic profit would be zero.

In summary, we can say producers will offer their product for sale if the product price is high enough to cover average production costs. The price at which a good will be supplied depends on the additional cost it takes to produce each additional unit of this good. In more general terms, that means that the supply curve is equal to the portion of the marginal cost curve (MC) above average cost (AC). Production theory distinguishes between supply in the short term and in the long run. The assumption is that in the short run, producers may be willing to have only their average variable costs covered and take some loss, due to the fact that closing down altogether would mean the loss of all fixed cost investments. In the long run, however, production can only be maintained if average fixed and average variable costs are covered. The long-run supply curve is, therefore, the marginal cost curve (MC) above average total cost. As in demand theory, we get from the individual producers' supply to the market supply curve for a commodity by simply adding up all the individual producers' supply levels at each corresponding product price level.

THE MODEL OF PERFECT COMPETITION

What is this perfectly competitive market we have repeatedly referred to? Perfect competition is the heart and soul of neoclassical analysis and policy recommendations. A perfectly competitive economy is one in which prices correctly reflect consumer preferences and production costs, and thus the conditions for general Pareto optimality established in Chapter 4 will be achieved if there is no interference with the market mechanism. The model of perfect competition describes an ideal type of *market structure*. By market structure economists mean the characteristics of markets in which goods are bought and sold. They usually include the number of buyers and sellers, the characteristics of the product, the relationship between the different firms in the market, and the conditions for entry and exit into and out of the market. The most common market structures in economic theory are *perfect competition, monopoly,* and *imperfectly competitive markets.*

A monopoly is characterized by only one firm producing and selling a particular good. If we had only one Brazil nut producer and no others offering a close substitute for Brazil nuts, we would have a monopoly situation in the Brazil nut market. Imperfectly competitive markets are described by a number of different models. An *oligopoly* is an industry that has only a handful of firms producing a similar product. They watch their competitors' actions and act accordingly. A market structure that is closer to perfect competition is *monopolistic competition.* It is characterized by a large number of firms producing similar but slightly different products. Other kinds of markets may be described as having a market leader and a competitive fringe. A variety of models exist to describe these and other kinds of market structures.

A perfectly competitive market has four important characteristics, all relating to price:

THE CHARACTERISTICS OF PERFECT COMPETITION

1. THERE IS A VERY LARGE NUMBER OF BUYERS AND SELLERS SO THAT NO ONE HAS THE ABILITY TO INFLUENCE PRICES.

2. THERE IS PERFECT INFORMATION, FREELY AVAILABLE TO ALL, ABOUT THE CHARACTERISTICS OF THE GOODS OR INPUTS BOUGHT AND SOLD, AND ABOUT THEIR PRICES.

3. THERE ARE NO MARKET BARRIERS. FIRMS AND CONSUMERS ARE PERFECTLY MOBILE AND CAN ENTER AND EXIT MARKETS EASILY SO THAT PRODUCTIVE INPUTS CAN BE TRANSFERRED FROM THE PRODUCTION OF ONE GOOD TO THE PRODUCTION OF ANOTHER WITHOUT COST.

4. ALL FIRMS WITHIN A PARTICULAR INDUSTRY PRODUCE IDENTICAL PRODUCTS, SO THERE IS NO REASON FOR CONSUMERS TO BUY ONE GOOD RATHER THAN ANOTHER EXCEPT WHEN PRICE DIFFERENCES EXIST.

Using the supply and demand curve model, we can now see how market interaction between consumers (demand) and producers (supply) establishes the equilibrium price. By supply and demand we mean the market supply and market demand curves that are made up of a large number of consumers and firms. The assumption is that we arrived at the supply and demand curves for beef depicted in Figure 5.8 by adding consumers' quantities of beef demanded and producers' quantities of beef supplied at each corresponding price level.

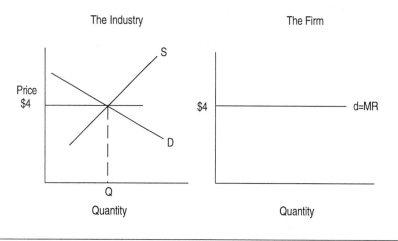

Figure 5.8 Competitive Markets and Firms.

In this case the equilibrium price for beef is $4. As a large number of buyers and sellers communicate their preferences in the market by deciding how much to buy or sell at a given price level, a level of "agreement" is established. At that agreed-on price, supply and demand are exactly equal. If for some reason the price is higher than $4, producers will want to sell more beef than what consumers are willing to buy. Supply will exceed demand driving the price lower. If the price is lower than $4, consumers will want to buy more beef than what producers are willing to offer at that price. Demand will exceed supply and the price will be driven higher. At a price of exactly $4, a stable or equilibrium situation is reached. This $4 price is called the *market clearing price*.

This equilibrium price is "given" to the firm. Given the assumptions of the perfect competition model, there is no reason for the firm to charge a lower price than the market clearing price. It would mean a loss in revenue. And since all goods are assumed to be the same and consumers are fully informed and fully mobile, they would buy their beef somewhere else if some producer tried to raise the prices for beef. If a firm raised its price even by a small amount, sales would drop to zero. Why would consumers pay $5 or even $4.01 when they can get an identical product for $4?

This means that the demand curve for the firm (d) is a horizontal line as shown in Figure 5.8. The firm is a price taker, meaning that any increase in price will cause sales to drop to zero given the four characteristics of perfect competition listed above. This is why we assumed earlier that for the perfectly competitive firm, marginal revenue (MR) is equal to price.

EFFICIENCY IN RESOURCE USE: LONG-RUN COMPETITIVE EQUILIBRIUM

Figure 5.9 depicts one of the most important concepts in economic theory—long-run competitive equilibrium. We saw above that a firm producing under the conditions of perfect competition can sell as much as it wants at the market clearing price.

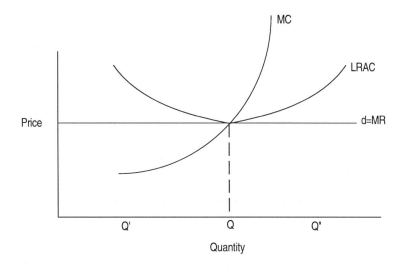

Figure 5.9 The Firm in Long-Run Competitive Equilibrium.

As shown in Figure 5.9, the level of output (Q) supplied in this market is where marginal costs are equal to the long-run market clearing price the producers face. We can also say that long-run marginal costs (MC) are equal to marginal revenue (MR). If firms produce less than Q (say 5,000 tons of beef), for example Q', marginal revenue is greater than marginal cost, and producers can increase profits by increasing output. This process of increasing profit by increasing production continues until point Q is reached. Now suppose output is higher than Q, at a point like Q". In this case profits can be increased by decreasing output since the additional costs generated at higher production levels exceed additional revenues. Any movement away from Q will reduce profit. At this point marginal cost is equal to marginal revenue (MC = MR), and price equals marginal cost (P = MC). At output level Q, producers earn "zero" economic profits. This means that there are no better (that is, more profitable) alternatives for the inputs to be invested than in this production process.

Another important piece of information characteristic of the long-run profit maximizing output level is that long-run average production costs are at their minimum. This indicates that the firm is producing its output at the lowest possible cost per unit, given society's resource endowments and available technology. The reason for this efficiency again follows from the assumptions of the perfect competition model. If a firm producing beef did not use the most efficient production method available, their marginal cost of production would be higher than other firms!. Since firms supply where MC = MR = P, others would be able to offer their product at a lower price and drive the less efficient firm out of business. As firms, however, leave production, the market supply for beef is reduced. The loss of beef producers causes beef supply to decrease, causing a shift of the supply curve to the left. As beef production drops and supply is lower than demand, beef prices are driven up. The price increase results in economic profits exceeding the normal "zero" profit level. This is a signal to potential producers that it is lucrative to either start or expand beef production. As beef production is increased, beef prices are driven down until the above-zero profits are eliminated, and we return to a level of zero economic profits. The main results of long-run competitive equilibrium are these (see next box):

> ### CHARACTERISTICS OF THE LONG-RUN
> ### COMPETITIVE EQUILIBRIUM
> 1. FIRMS OPERATE AT MAXIMUM EFFICIENCY, THAT IS, AT THE LOWEST POSSIBLE PER UNIT COST (AVERAGE COST).
> 2. THE PRICE OF THE GOOD PRODUCED WILL BE EQUAL TO ITS MARGINAL COST OF PRODUCTION.
> 3. THE FIRM WILL EARN A ZERO ECONOMIC PROFIT, THAT IS, THE RATE OF PROFIT WILL BE EQUAL TO THE PROFIT RATE PREVAILING IN THE ECONOMY.

PERFECT COMPETITION AND PARETO OPTIMALITY

At the beginning of this chapter we found that the marginal rate of substitution of good X for good Y (beef for Brazil nuts) is equal to the ratio of their product prices: $MRS = \Delta X / \Delta Y = (\Delta U / \Delta Y) \div (\Delta U / \Delta X) = P_Y / P_X$. This holds for all consumers, $MRS^{XY}_{Bertha} = MRS^{XY}_{Alex}$.

Likewise, from our discussion of production under a cost constraint, we found that the marginal rate of technical substitution of inputs (in our example, Labor for Capital) is equal to the ratio of their input prices: $MRTS = \Delta K / \Delta L = (\Delta Q / \Delta L) \div (\Delta Q / \Delta K) = P_L / P_K$. This holds for all goods produced. $MRTS^{KL}_{beef} = MRTS^{KL}_{Brazil\ nuts}$.

From our supply curve discussion we found that the supply of a product is determined by the marginal costs of its production. A product's supply curve is its marginal cost curve above average production costs. Since profits are maximized when marginal costs are equal to marginal revenue and producer's marginal revenue is given by the market clearing price, marginal cost is equal to product price for all goods produced under perfect competition:

$MC_X = P_X$ and $MC_Y = P_Y$.

We found further that under the conditions of perfect competition, inputs are used at maximum efficiency since a market's long-term competitive equilibrium is reached at the lowest average production cost level possible. Thus the long-term marginal costs of producing good X (MC$_X$) and the long-term marginal costs of producing good Y (MC$_Y$) give us the rate of product transformation of goods X and Y, that is, the combinations of X and Y that can be produced if inputs are used as efficiently as possible.

With this we have established Pareto optimality under perfect competition:

PARETO OPTIMALITY UNDER PERFECT COMPETITION

1. MRS$_{Y FOR X}$ = P$_Y$/P$_X$. THE MARGINAL RATE OF SUBSTITUTION OF GOOD Y FOR GOOD X IS EQUAL TO THE RATIO OF THE PRODUCT PRICES OF THESE GOODS FOR ALL CONSUMERS.

2. MRTS$_{L FOR K}$ = P$_L$/P$_K$. THE MARGINAL RATE OF TECHNICAL SUBSTITUTION OF CAPITAL INPUTS FOR LABOR INPUTS IS EQUAL TO THE RATIO OF THE PRICES OF THESE TWO INPUTS FOR ALL GOODS PRODUCED.

3. MC$_Y$/MC$_X$ = P$_Y$/P$_X$ = RPT. SINCE MRS = P$_Y$/P$_X$, IT FOLLOWS THAT MRS = RPT, THE RATIO OF THE MARGINAL PRODUCTION COSTS OF TWO GOODS X AND Y IS EQUAL TO THEIR PRODUCT PRICE RATIO.

We have replicated the conditions of general equilibrium established in Chapter 4 for a society operating under the conditions of perfect competition. In this society the price system takes on the function of communicating consumption preferences and production efficiency.

PRICES AND THE BIOPHYSICAL WORLD

The introduction of prices into the model of Pareto optimality makes this model more realistic, since there is no longer a need for consumers and producers to interact in a face to face manner; but prices also add to the problems we discussed earlier. All the limitations we saw in the first four chapters, including the problem of pure time preference, the lack of information about qualitative difference between goods that affect environmental resources, and the presence of threshold effects and irreversibilities, are carried over when we add prices to the neoclassical general equilibrium model.

But we also introduced another set of complications. First, the assumption that the economy is characterized by perfect competition is an obvious leap of faith. However, we will not discuss the objections to these assumptions in this book since (1) our objective is to limit our critique to the theoretical limitations of the general equilibrium model, and (2) the limitations of the model of perfect competition have been extensively discussed elsewhere. Most of the criticism of economic theory by environmentalists has focused on the unrealistic assumptions of the competitive equilibrium model (see the references for Henderson and Sagoff at the end of this chapter).

Second, for perfect competition to lead to Pareto optimality, the price signals sent through the economy must correctly reflect not only individual values, but social values. When it comes to irreplaceable resources, irreversible effects, or irreducible pollution, this is an impossible task for the market. Values have to be communicated by market participants or their influence will not be felt in the market. Yet marginalized social groups, future generations, and other species, which are or will be affected by today's environmental degradation, have little or no opportunity to bid in the auctioneer's market to influence market prices according to their preference. The economist Nicholas Georgescu-Roegen points out that putting prices on irreplaceable natural resources is like auctioning off the Mona Lisa to the students in a single classroom. The price of this painting would be ridiculously low since other interested parties cannot bid. The problem is insurmountable, even theoretically, since, if present and future generations most strongly affected by our collective economic activities could bid for such resources as rainforests, the protective ozone layer, water, and air, their price would be enormously high. If they are excluded from bidding, the market is imperfect since not all have equal access, information, and voice. Suppose the present generation could go back in time and bid for now extinct animals such as the Tasmanian marsupial wolf, or the Great Auk, or the Dodo, or plants that might lead to cures for diseases like AIDS or cancer. The price offered would certainly be high enough to preserve these species. However, they were all driven to extinction in the past because the price was just right to kill one after the other.

The third limitation is the assumption that a single measurement, namely price, can reflect the myriad attributes, interrelationships, and functions of an environmental feature. Neoclassical theory assumes that all human wants can be reduced to a common denominator: price. Likewise, all resource values can be reduced to a common denominator: input price. This is a general problem with the market evaluation of any collection of goods, but it is a particularly difficult problem for the evaluation of environmental attributes. How do we evaluate "priceless" services and functions whose attributes are not only unknown but unknowable? The ecological and evolutionary value of biological species in insuring the long-run adaptability of life to the changing conditions of planet Earth can never be known. Yet neoclassical theory expects the market system to place a price on an individual species, or even on an ecosystem, which would then correctly reflect their market value to humans. Take the example of Brazil nuts. If Brazil nuts were discovered to be tremendously beneficial to human health, demand would most certainly increase, as would their price. If we assume that they cannot be produced in very large quantity, since they are dependent on tropical forest conditions, the deforestation that has taken place with increasing speed over the past three decades would certainly be recognized as a mistake. At the same time, however, the price increase for Brazil nuts may send a different message. Even though protection to secure future production would be recognized, the temptation to cut down the forests and intensify production in Brazil nut plantations would likely lead to rain forest destruction, not in spite of, but because of their high value.

The folly of the underlying assumption of universal substitutability in a Pareto optimal economy is particularly transparent when prices are introduced. Not only does neoclassical theory say that individual items are substitutable for each other, but they can be substituted by their dollar equivalents. With this reasoning it is permissible to cut down a rainforest, for example, as long as we put the money in the bank where it is available for future investment. And it is acceptable to destroy "natural capital" as long as its destruction is offset by an equivalent increase in savings. This is the so-called "weak sustainability" criterion of neoclassical-classical economics. We believe that the insurmountable problem is not so much that market prices are incorrect, but that prices in and of themselves are incapable of reflecting environmental values.

Neoclassical economists recognize that prices in the private market may not reflect true social or ecological values and that market failure may be present. The consequences of such market failures are discussed in Chapter 6.

SUMMARY

The two major ideas presented so far are the general equilibrium conditions of Pareto optimality in a pure exchange system (Chapters 2, 3, and 4) and the demonstration that a perfectly operating price system will achieve the same result (this chapter). The third major principle of neoclassical theory is the idea of market failure. Perfect competition will insure Pareto optimality only if the proper price signals are sent. Economists recognize that this is not always the case. Sometimes the market fails to send the proper price signals to consumers and producers, and so Pareto optimality is not achieved. Market failure is discussed in the next chapter.

Most of the criticisms of the neoclassical theory of perfect competition have focused on the unrealistic underlying assumptions of the model. Such criticisms are relevant, but they miss two important points: (1) even if all the assumptions of the neoclassical theory were true, it would still be incompatible with environmental sustainability because of discounting, irreversibility, and the impossibility of placing a single price on environmental attributes; and (2) in many ways, the competitive model is an accurate, if idealized, description of real-world markets. The reasons for the unsustainability of market economies are made clear by the neoclassical model of market exchange.

SUGGESTIONS FOR FURTHER READING

Georgescu-Roegen, Nicholas. *Energy and Economic Myths*. Pergamon Press, New York, 1976.

Gowdy, John and Peg Olsen. "Further Problems with Neoclassical Environmental Economics," *Environmental Ethics* 16 (1994), 161–175.

Henderson, Hazel. *Creating Alternative Futures*. Berkley Press, New York, 1978.

Norton, Brian. "Thoreau's Insect Analogies: or Why Environmentalists Hate Mainstream Economists," *Environmental Ethics* 13 (1991), 235–251.

Sagoff, Mark. *The Economy of the Earth*. Cambridge Univ. Press, New York, 1988.

6

MARKET FAILURE: WHEN PRICES ARE WRONG

INTRODUCTION

The function of prices in the neoclassical model is to send signals through markets that tell consumers and producers the characteristics of market goods and productive inputs. As discussed in Chapter 5, a perfectly operating price system will lead to Pareto optimality as described in Chapters 2, 3, and 4. In real world situations, however, there are a variety of reasons why prices might not send correct signals about the characteristics of goods and services to consumers or about the characteristics of inputs to producers. In these cases, not only the collective but also the individual preferences of market participants are distorted, and Pareto optimality is not achieved. Neoclassical economists recognize these shortcomings and refer to them as *market failure*. We saw in the last chapter that the condition for a perfectly operating market economy to reach Pareto optimality is that market prices correctly reflect consumer preferences, input prices correctly reflect productivity, and the prices of goods must equal their marginal cost of production. These conditions are violated when market failure is present. Three general types of market failure relevant to environmental issues are (1) imperfect market structure, (2) public goods, and (3) externalities.

Standard economic policies to correct market failure concentrate on establishing the conditions of Pareto optimality. Neoclassical environmental policy, therefore, begins (and usually ends) with recommendations to adjust relative prices. In addition to price adjustments, policy intervention measures include regulation or voluntary compliance. As policymakers seek to correct the value the price system assigns to the functions and services of our biophysical environment, they face such challenges as intervention failure and existence failure. These refer to the fact that policy measures themselves may send erroneous signals or fail to assess the value of environmental goods and functions beyond usefulness to humans as defined through market exchange (see Figure 6.1).

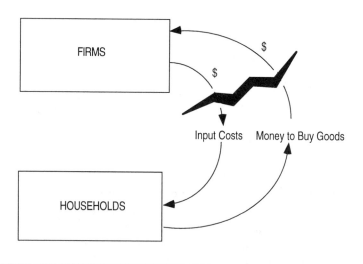

Figure 6.1 The Effect of Market Failure on Market Exchange.

IMPERFECT MARKET STRUCTURES

From the outset our focus has been on the ideal model of perfect competition, so we will only offer a brief discussion of the effects of suboptimal market structures such as monopolies, oligopolies, or monopolistic competition. Unlike the case of perfect competition, in all these imperfect market structures a product's marginal revenue is different from the market clearing price. In the perfect competition model discussed in Chapter 5, we saw that the market clearing price was established by the interaction between supply and demand in the entire industry. This price determines the additional revenue (MR) each individual firm receives for each additional unit of output sold, so that individual firms cannot influence the product price. In the case of imperfect market structures, firms have pricing power and, therefore, do not have to accept the market clearing price. Imperfectly competitive firms can increase or decrease sales by lowering or raising the price. This means that each firm faces a downward sloping marginal revenue curve. As firms decrease production, marginal revenues increase; as production is increased, marginal revenues decrease.

The most extreme case is that of a monopoly, defined as a single producer of a product for which there are no close substitutes. Monopolies may come about as a result of (1) one firm having control of an essential productive input, (2) having increasingly lower per unit costs of production as the level of output increases (this is called a *natural monopoly*), and (3) a firm owning patents and franchises that protect it from competition. But why should monopoly power be a problem? Figure 6.2 illustrates some of the consequences of monopoly power compared to perfect competition. If this industry was competitive, output would be determined by the intersection of the demand curve, representing consumers' preferences, and the supply curve, representing producers' marginal costs when profits are maximized. A competitive industry would produce an output of X_C and the price of the good would be P_C. If a monopoly were the sole producer in this market, an output level of X_M would be produced, and the market price charged would be P_M. According to neoclassical theory, therefore, monopolists will charge a higher price and produce a lower level of output.

An important concept related to Figure 6.2 is consumer surplus. Suppose that the market for beef is competitive and the price is P_C. If the price was higher, would there still be some demand for beef? The answer is yes, all the way up to point A in Figure 6.2. At that point the price is so high that the demand for beef would be zero. Since the market price is P_C, this means that some people would be willing to pay more than the market clearing price (between P_C and A).

> **THE DIFFERENCE BETWEEN WHAT CONSUMERS ACTUALLY PAY AND WHAT THEY ARE WILLING TO PAY IS *CONSUMER SURPLUS*.**

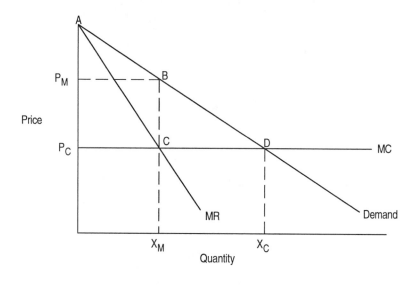

Figure 6.2 The Social Cost of Monopoly.

In the case of perfect competition, the total amount of consumer surplus is the triangle AP_CD. If the industry were a monopoly, consumer surplus would be reduced to the area AP_MB. Part of the loss in consumer surplus is simply a transfer to producers, who would gain the area P_MBCP_C in the monopoly situation. This transfer is not considered a loss to society since one group gains from another group's loss. But some consumer surplus equal to the triangle-shaped area BCD is lost altogether. This is called *deadweight loss* and is a measure of the welfare loss to society due to the imperfect market structure. This loss is due to the fact that monopolies produce a lower level of output than what is socially desired.

The notion of monopolies "underproducing" has come under fire from critics of the neoclassical model like John Kenneth Galbraith. He argues that the real problem with monopoly power is not underproduction but overproduction, and that sectors characterized by market power, such as the energy sector and the transportation sector, produce too many private goods and not enough public goods. For example, too much is spent on private automobiles and too little on public transportation. Galbraith argues that, through advertising, producers are able to create a demand for their products in excess of the social optimum.

In addition, monopolies do not necessarily produce at the minimum long-run average cost. Thus the most efficient use of inputs that characterizes perfectly competitive industries is not guaranteed. In regulating monopolies, such as public utilities, regulators often try to duplicate the results of competition by imposing *marginal cost pricing*, that is, utility rates are set as close as possible to the marginal costs of providing service.

Monopolistic Competition markets are characterized by a large number of producers with similar but differentiated products. In this market structure, output is not produced at the lowest possible per unit cost of production. The result is *overcapacity*. Such firms could increase output, but maintain overcapacity to differentiate their product from their competitors.

Because imperfectly competitive industries produce a lower output than they would under perfect competition, some economists have called monopoly power "the conservationist's best friend." If the structure of the petroleum industry or the energy sector, for example, was one of perfect competition rather than oligopoly, annual output would be higher, oil prices lower, and the resource would be used up more quickly. However, particularly in oligopoly markets, competition between firms (as in the petroleum sector, for example), is likely to work against rather than for resource conservation as firms try to capture market shares. There is no guarantee that social goals of conservation will be upheld or supported by monopolistic markets. It is more likely that socially motivated influences on private production and resource use decisions are reduced as monopolistic power increases.

PUBLIC GOODS

Public goods have two characteristics that make it impossible for them to be accurately evaluated by the price system—they are *non-exclusive* and *non-rival*. Non-exclusive means that once the good is provided, people can use it whether they are willing to pay for it or not. Non-rival means that one person's use of the good does not preclude others from using it. Public goods are distinct from the private market goods described in Chapter 5. Private goods or services are rival and exclusive. If consumers want to buy a good (beef, for example), they have to pay for it. No money, no beef. And if ten pounds of beef are sold, there is less for others. One person's use precludes another person from using the good who cannot eat the beef without the permission of the owner (exclusive). For public goods, such as a public radio station, no one can be excluded from listening once the program is on the air, whether they send money to the station or not. This is called the *free rider* problem of a non-rival good. Additional people can turn on their radios and listen to the program without affecting any other user. Since there are no proper price signals to reflect the demand for pubic goods, the market has no way of getting information about how much of the good to provide. Recall the necessary condition for Pareto optimality under perfect competition—namely, that the price of a good has to equal its marginal production cost. Public goods, once they are provided, have zero marginal costs.

Not all goods are either purely private or purely public. National parks, for example, tend to be exclusive but non-rival. Entrance to these parks can be controlled by charging a fee for their use (exclusion); one more person's use will not detract from any other person's enjoyment (non-rival). But as more and more people visit the park (such as Yellowstone National Park in July), its use will become "rival," and some mechanism may have to be used to limit the number of people in the park, either by charging higher fees or through a lottery system. The classic case of a fishing ground without entrance restriction or fees is an example for rival but non-exclusive goods. Access is free, but too many people become an impediment to fishing (rival). Table 6.1 shows all four cases of private, public, and mixed goods.

Table 6.1 Characteristics of Public Goods

	Rival	Non-Rival
Exclusive	Pure Private Goods	Private Beach (uncrowded)
Non-Exclusive	Open Access Fishing Ground	Pure Public Goods (Public TV)

Rival but non-exclusive goods have a long and confused history in environmental economics. Neoclassical economists see a continuum ranging from private property under the exclusive control of an owner to what has been erroneously called *common property*, which is owned by no one. If anyone can go to the fishing ground and catch fish (non-exclusive), there is insurmountable pressure to over fish and eventually exhaust the resource (rival). In the neoclassical literature dealing with public goods, everything from over exploitation during the early fur trade in Canada to the current loss of biodiversity and genetic resources is blamed on a lack of private property. This argument was spurred by Garret Hardin in his influential article, "The Tragedy of the Commons" where he argued that environmental destruction has been the result of over-exploitation of natural resources due to a lack of private ownership. This analysis leads to an easy solution that involves minimal government involvement and encourages an expansion of the private markets—namely, to assign private property rights.

The problem with this approach is that it confuses "common property" with a situation in which resource use is unregulated. Access to the original medieval commons, for example, was regulated by social customs to prevent over-exploitation. Native Americans, as is now well-known, had elaborate rules and religious customs which had the effect of guarding against over-hunting, at least until the advent of the European fur trade and the introduction of market goods. A more useful distinction, therefore, is between "common property" (controlled by some communal or social rules) and *open access* (unrestricted use). The neoclassical literature confuses the two by considering only the case of unrestricted use. This sets up an unrealistic continuum between private property and open access resources. With only these two cases it is easy to prove that assigning private ownership will result in a better allocation of resources and lead to the optimal amount of conservation. Neoclassical analysis, with its emphasis on individual decision-makers, precludes the consideration of common property situations where shared responsibility promotes a more rational use of resources in the long run. This goes back to the Edgeworth box analysis described in Chapters 2, 3, and 4. There is no place for society's preferences to be communicated. Only individual agents striving to maximize utility in a system of static exchange are considered. In open access cases, resources are used inefficiently because they are not properly assigned to an owner to give signals to the market that would reflect their worth relative to other goods. Assigning property rights returns us to the system of exchange described by the Edgeworth box diagram.

EXTERNALITIES

The last case of market failure is that of externalities. Economists speak of negative externalities when (1) the welfare of one consumer or producer is adversely affected by the actions of another, and (2) this loss of welfare is uncompensated. Following the economic division between producers and consumers, there are four kinds of externalities: (1) externalities between firms, (2) between consumers, (3) externalities imposed by consumers on firms, and (4) those imposed by firms on consumers. Examples include a firm producing paper products whose sludge pollutes the brook supplying water to the cattle ranch down stream; or a consumer driving a car to make a purchase at the store and emitting NO_x etc. in the process; or the beef producer who grows silage corn for feed and fertilizes it with such high levels of manure that in neighboring wells, drinking water standards for nitrates are violated. We will limit our theoretical discussion to the fourth type: producers affecting consumers. In this case the *private* marginal cost of producing a product (and thus its price) does not reflect its *social* cost.

Figure 6.3 shows that the social marginal cost of producing beef (MC_S) is higher than it's private marginal production cost (MC_P). The profit maximizing level of output from the producer's point of view is determined by the intersection of the marginal revenue and the marginal cost curve, and consequently X_P would be produced. The social costs of beef production, however, are higher than the labor, fuel, feed costs, etc. the producer has to pay. They include the costs resulting from soil erosion, water pollution, biodiversity loss or whatever other costs might occur in a particular case. If the producer had to pay these costs to society, total marginal costs would be higher (shown by the marginal cost curve MC_S), and output would drop to X_S and the product price would increase from P_P to P_S. From this simple diagram, we can see the two major characteristics of a situation in which negative externalities are present; (1) output is too high, and (2) the price of the good is too low. Again, the main concern of neoclassical economic analysis is to determine a product's price relative to other products to see if these prices insure that Pareto optimality is achieved. Environmental policy recommendations seek to include the costs of environmental damages caused by the production of a good in the price of this product so that the market will allocate society's resources correctly. This is called *internalizing the externality*.

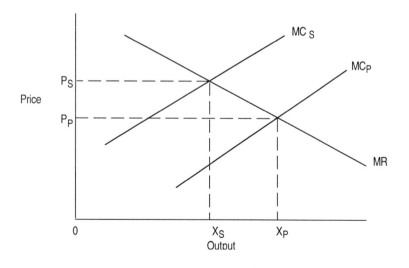

Figure 6.3 Production Cost with Externality.

As we have seen, a basic assumption of neoclassical theory is that everything is substitutable, can be traded, and expressed in monetary units. Consequently, there is a trade-off between pollution and consumer goods whose production causes pollution. We value both consumer goods and environmental quality, for example, both beef and the hydrological service of undisturbed forest land. We choose a combination of environmental protection and the production that maximizes our utility. This means that there is a "socially optimal" amount of pollution that is greater than zero. Figure 6.4 illustrates this notion of the optimal amount of pollution. The benefits from allowing pollution to take place are characterized by the MNPB line (which stands for *marginal net private benefit*). It shows the marginal benefits from producing good X (beef). The costs associated with the pollution generated are shown by the MSC, or *marginal social cost* line. If no consideration is given to the cost of pollution, an output level of X_P would be produced. The total private benefit would be the area under the line MNPB, or A+B+C. The social cost of the pollution generated by this level of output would be the area under the MSC line or B+C+D.

The optimal level of production is X_S where marginal net private benefits are equal to marginal social costs. At any amount less

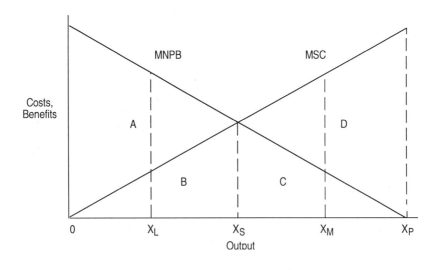

Figure 6.4 The Optimal Level of Pollution.

than X_S, say an amount X_L, an increase in output adds more benefits than costs to this society. At any point to the right of X_S, say at an output level of X_M, the social costs of additional production are greater than the benefits. At X_M if we reduce production, we are reducing costs more than benefits and thus a reduction in output is positive for society. At output level X_S, an amount of pollution equal to the area C + D has been eliminated. It is important to note, however, that costs and benefits are considered here as monetary amounts. Distributional issues or considerations about who suffers from the damages or who benefits do not enter this concept of optimal pollution. It is further assumed that the marginal social costs of pollution can actually be determined and related to a specific level of production (for a concrete example, see Chapter 8).

SOLUTIONS TO THE EXTERNALITY PROBLEM—COASE VERSUS PIGOU

The most commonly proposed solutions to the problem of negative externalities are (1) assigning property rights, and (2) taxing the polluter. According to neoclassical theory, both solutions insure that the optimization point will be reached where the marginal cost of pollution equals its marginal benefit.

The property rights solution states that if property rights are assigned to things like air and water, then Pareto optimality will be reached through negotiation between the parties involved. Take the example of the beef farmer who grows silage corn and contributes to the pollution of nearby residents' well water. If property rights to the watershed are assigned to the producer, residents would pay the farm not to pollute the water. If property rights are assigned to the residents, the farm would pay them for the right to pollute. In either case, if there are no transactions costs, that is, no costs of negotiating and monitoring the agreement, then the end result will be the optimal amount of pollution generated by output level X_S, as shown in Figure 6.4. This property rights solution is called the *Coase theorem*, named after the economist Ronald Coase. Its assumption is that the private market will automatically take care of the externality problem. The major drawbacks to this solution are its assumption that perfect information about the true costs of pollution actually exists, that transaction costs are negligible, and its limited feasibility in real world situations where it is generally exceedingly difficult to bring affected parties together to negotiate pollution rights and prices.

The Pigouvian solution, named after the economist A.C. Pigou, is to tax the polluter an amount equal to the external cost. Referring back to Figure 6.3. the idea of this solution is to set the per unit tax so that the marginal cost of production to the producer is equal to the marginal social cost. This tax would be equal to the difference between MC_P and MC_S in Figure 6.3. Referring to Figure 6.4, by increasing marginal production costs, the Pigouvian tax would reduce the marginal net private benefit and shift the MNPB curve downwards until it intersects the marginal social cost curve (MSC) at the optimal output level X_S. The effect of the tax would be to lower output (thus reducing pollution) and raise the price of the good.

Other measures that would have the effect of increasing the marginal cost of production are regulations requiring specific production methods or technologies. Again, such solutions assume that the amount of pollution generated at various levels of production and with various production methods can be determined and that the marginal social costs of pollution (and its corresponding production level) are known. They eliminate, however, the need for negotiation between the polluter and those affected by the pollution.

ELASTICITIES—MEASURING POLICY EFFECTIVENESS

A question we have not addressed so far is who pays for an intervention measure like the Pigouvian tax? While it might seem that the polluter pays, this is not necessarily so since producers will try to pass at least some of the increased marginal production costs on to consumers. One of the most widely used economic tools to analyze the effects of policy measures is the measure of elasticity. *Price elasticities* measure the effect of a small change in price on the quantity of a good demanded or supplied. Other elasticities might seek to measure the impact of a change in income (*income elastici-*

ties) or a change in the price of a related good (*cross price elasticity*) on the quantity demanded of a good.

The demand curve we introduced in Chapter 5 is called an ordinary or *Marshallian* demand curve (named for the British economist Alfred Marshall). It can serve as the basis for the concept of price elasticity. The demand curve shows the responsiveness of consumer demand to price changes as the change in quantity divided by the change in price. For good X (beef) we get ($\Delta X / \Delta P$), which is the inverse of the demand curve slope ($\Delta P / \Delta X$). For elasticities we use percentage changes rather than absolute changes. The *price elasticity of demand* (ε_X) is defined as the *percentage* change in the quantity of beef demanded divided by the *percentage* change in its price. This can be written as:

$$\varepsilon_X = (\%\Delta X) \div (\%\Delta P_X) = (\Delta X / X) \div (\Delta P_X / P_X) = (\Delta X / \Delta P_X) \bullet (P_X / X).$$

Because of the law of demand, which states that price increases lead to a decrease in the quantity demanded and vise versa, the price elasticity of demand is always less than zero. An elasticity of –0.5 for beef indicates that a 2 percent increase in the beef price would cause the quantity of beef demanded to fall by 1 percent. If the absolute value of the elasticity is less than 1 ($0 < |\varepsilon_X| < 1$), the demand for the good is said to be *inelastic* or not very price responsive. If the elasticity of a good is greater than 1 in absolute value ($|\varepsilon_X| > 1$), demand is said to be *price elastic*. If a highly elastic good is taxed, consumers are likely to respond to the price increase by drastically reducing their purchase. Thus producers would absorb a large part of the price increase. For inelastic goods such as food or other necessities, consumers would not reduce their purchases very much despite increases in price and thus would absorb the main part of the tax.

Different elasticities on both the supply and demand side lead to very different levels of responsiveness to policy measures. Consider, for example, the implementation of an energy tax to reduce the consumption of fossil fuels. Since energy is a necessary input in production and consumption, items such as gasoline and electricity have low price elasticities of demand, around –0.3 or –0.4. This means that a relatively high tax (as a percentage of per unit energy costs) would be necessary to significantly reduce consumption.

However, if the purpose of the tax is to raise revenue, such a tax would be effective since consumption of the good (and therefore tax collections) would remain high even if the tax raised prices significantly.

Cross-price elasticities can also provide important information about environmental policies. For example, the cross price elasticity would indicate how an increase in the beef price would impact Brazil nut consumption. Assuming that not many consumers would substitute their hamburgers for roasted Brazil nuts, the cross-price elasticity for these two goods would likely be low and positive. A positive cross price elasticity indicates a certain degree of substitutability between two goods as opposed to complementarity, which exists for such goods as hamburgers and hamburger buns that are used together. Complements have a negative cross-price elasticity. For example, as the price of beef patties goes up, the demand for hamburger buns goes down. The higher the absolute value of the cross-price elasticity, the higher the degree of substitutability or complementarity between two goods. Using taxes or regulations, or consumer boycotts to reduce the consumption of an environmentally harmful product will be more effective (and more politically acceptable) if there are close substitutes for that product.

The *elasticity of supply* shows how much producers (sellers) will change the amount of a good supplied (X_S) in response to a change in the good's price. In the debates during the 1970s and 1980s over decontrolling energy prices, the argument in favor of decontrol was that higher prices were necessary to call forth a greater supply of natural gas and petroleum. Opponents pointed out that the elasticity of supply for these fuels is relatively low and thus higher prices would not have a great effect on the quantity supplied.

There are also elasticities associated with the firm's production decisions. These are formally identical to the ones discussed above. Elasticities of factor demand show the percentage change in the quantity demanded of a factor of production (land, labor, and capital) resulting from a percentage change in the price of that factor. The most widely used cross-price relationship in production is the *elasticity of substitution*. It has been widely used by economists to estimate the ability of the economy to adjust to increasing natural resource prices by substituting other inputs. One widely debated example is the question of whether capital and energy are substitutes or complements in production. If they are substitutes, then increasing energy scarcity can be partially offset by using capital instead. If they are complements, then increasing energy scarcity and higher energy prices will result in a lower rate of capital formation and a slowdown in productivity growth.

CONSUMER SURPLUS—WHO PAYS FOR INTERVENTION?

As mentioned in our discussion of imperfect market structures, consumer surplus is another concept that addresses the question of who is affected by policy measures and in what way. It is widely used as a measure of welfare change. The application of the concept seems straightforward, but there are two important complications in this analysis. First, as we have seen, price changes cause changes in buying power, and thus in real income. Second, in empirical estimates of consumer surplus, it seems to matter whether we are measuring the welfare loss from a price increase, or the welfare gain from a price decrease.

The first complication arises from the income effect mentioned in Chapter 5. Whenever there is a price change in a good without a change in budget, there is a change in real income. To account for this, economists use a "Hicksian" demand curve (named for the British economist J.R. Hicks). Recall that the Marshallian demand curve assumes that money income (nominal income) is constant and only the price of one good is allowed to vary. The Hicksian demand curve also shows the relationship between a good's price and the quantity demanded, but it holds real income constant by holding utility constant. Recall from Chapter 5 that if there is an increase in the price of Brazil nuts, the effect will be a decrease in the quantity of Brazil nuts demanded. This change is made up of two components, the substitution effect and the income effect. If we assume that we can compensate our consumer Bertha for the price increase by giving her enough income so she can stay on her original indifference curve then we have eliminated the income effect and isolated the substitution effect. The income-compensated Hicksian demand curve is more steeply sloped (less price responsive) than the ordinary demand curve, because it shows only that part of the price responsiveness of demand which results from the substitution effect. In measuring consumer surplus, it makes a difference whether or not we hold real income constant when prices change. Because of the real income effect, there is a difference between how consumers react to a price increase or a price decrease of the same magnitude.

Measures of consumer surplus are used by economists to calculate the economic value of environmental goods. This is based on the idea that if one determines how much consumers would be willing to pay (WTP) to maintain or improve the quality of an environmental good, or how much they would have to be compensated in order to be willing to accept the loss or deterioration of an environmental good (WTA), one gets an estimate of the value of the environmental good in question. Empirical studies based on questionnaires, called *contingent valuation* (CV), show that WTA measures are much higher than WTP measures. There is an ongoing debate as to why this is so. It may be simply that individuals value the loss of something they already have more than they value the gain of something they do not have. It may also be that the discrepancy can be partially explained by the differences between the Hicksian and the Marshallian measures discussed above. Since the Marshallian demand curve includes the income effect and the Hicksian demand curve does not, one could speculate that the more income elastic a good is, the greater will be the effect of a real income change on quantity demanded and the greater will be the difference between the Hicksian and the Marshallian measure of consumer surplus; this would mean a larger difference between WTP and WTA measures.

In addition to income levels, the degree of information about the value of an environmental good, and the extent to which respondents are directly affected by changes in environmental quality, will result in variations between consumers' willingness to pay and willingness to accept.

Regardless of the differences between the empirical methods of contingent valuation, the method as such remains firmly embedded in an economic mind-set of individual preferences, market valuation, and monetarization. Critics note that this framework simply cannot reflect such critical functions as the protection of the life-supporting attributes of ecosystems, or the irreversibility of damages and concerns for intergenerational allocation needs. Indeed the reaction of many CV respondents, "But I can't say that. This is an entirely different kind of thing..." is indicative of doubts about the appropriateness of these valuation methods.

INTERVENTION FAILURE

Intervention failure is a type of distortion that occurs when public intervention in a market moves that market further away from a Pareto optimal position. Intervention failure has had detrimental effects on the environment and has diverted attention from the types of private market failure discussed above. It includes taxes and subsidies that affect prices, as well as regulatory measures such as technology or production specifications.

Figure 6.5 shows two government responses to the presence of a negative externality. Without government intervention, marginal cost would contain only private costs illustrated by the curve MC_P. The result would be an output of X_P tons of beef and a beef price of P_P. To eliminate the negative externalities of beef production short of taxing beef, one might seek to restrict intensive production methods by restricting the areas used for silage corn production or intensive grassland management. These measures would likely result in an increase in the marginal costs of beef production. Ideally an increase in marginal costs equal to the difference between private cost and social cost would be sought. The result would be the marginal cost curve MC_S yielding an output level of X_S, and a market price of P_S. This is the idea of "internalizing the externality" discussed in Chapter 4.

Intervention failure actually worsens the existing situation by moving the market to a marginal cost curve, such as MC_F which leads to an output level of X_F and a price of P_F. In this case the activity creating a socially undesirable environmental cost has been subsidized, causing a lower price and a higher output of the activity than in the private market case. To stay with our beef example, if government sought to compensate farmers for the income losses resulting from increased production costs by allowing the conversion of additional forests into farmland or paying subsidies to beef producers, the desired improvement in environmental quality would be undone, and the situation is likely to be worse than before (see also Chapter 8).

Unfortunately, intervention failure is pervasive, and perhaps inevitable, in societies with modern governments made up of competing bureaucracies. An example is the subsidized wetland conversion promoted by the U.S. Department of Agriculture prior to 1985. Most resource economists believe that wetlands are undervalued as a resource because their (non-market) role in flood control, as wildlife habitats, and for recreation and open space is not properly accounted for in the value they are assigned. Thus their destruction would be moving to the cost curve MC_F in Figure 6.5 further away from the social optimum than if left to the private market. As is common in the case of intervention failure, other U.S. government agencies such as the Interior Department pose direct opposition to the Department of Agriculture's initiative by providing incentives to protect and restore wetlands.

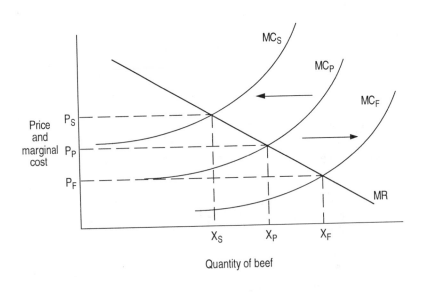

Figure 6.5 Intervention Failure

Intervention failure is also not limited to national government agencies. The World Bank has long been a target for financing such ecologically disastrous projects as the transmigration program in Indonesia, which resettled tens of thousands of residents from thickly populated Java to remote forest areas that were quickly cleared for settlements and agriculture. Other examples include roads to remote areas of Brazilian rainforests with disastrous results, and scores of dams in environmentally sensitive areas, all financed through World Bank funds.

In addition to government actions subsidizing environmental destruction, the problem with intervention failure is that many economists mistake *part* of the truth for the whole and claim that all environmental misuse is the result of misguided government policies. Advocates of this position fall back on the property rights argument discussed above and assert that by assigning property rights to all environmental goods—from whales to wolves to the atmosphere—all environmental problems can be solved. Recent history has made it clear that markets are the best system yet devised to allocate a given collection of resources. However, even if the allocation process was made economically optimal by correcting market failure and eliminating intervention failure, there is no guarantee that the economic optimum achieved is consistent with environmental sustainability.

EXISTENCE FAILURE

Existence failure refers to the inability of private markets to determine an optimal level of economic activity that adequately reflects the value of environmental goods beyond their market value. The result is a failure to preserve critical environmental functions. There is no guarantee that an economic optimum is consistent with maintaining the biophysical functions that ultimately sustain human activity. We may reach the goal of optimal allocation within the Edgeworth box framework, but this does not solve the problem of determining the size of the box. In formulating environmental policies, we need to distinguish between allocation and scale. Herman Daly uses the analogy of loading a boat to describe the neglect of scale in "optimal" market outcomes. The criteria of optimal market allocation insures that the boat is loaded evenly, but since nothing is said about the size of the load, it may also assure that the boat sinks "optimally."

Economists have proposed a number of measures that address this problem. Ciriacy-Wantrup and Bishop have advocated a Safe Minimum Standard approach to environmental policy. According to this approach, when there is a suspicion that actions may irreversibly harm environmental functions, we should always err on the side of caution. A similar concern is expressed in the Precautionary Principle referred to in *Agenda 21* prepared by the UN Conference on Environment and Development. This principle states that uncertainty with respect to the effects of human activities on the biophysical world cannot be used as an excuse for inaction in taking preventive and precautionary measures. One of the best statements of this position is by Nicholas Georgescu-Roegen, who said that the economic problem is not to maximize utility, but rather to minimize regrets.

As mentioned previously Contingent Valuation is another measure that seeks to place an existence value on environmental attributes and function. Despite the known problems with the neoclassical framework on which Contingent Valuation methods are based, this method has received wide attention. A noted CV study estimated the monetary loss of environmental quality to Americans from the Exxon Valdize accident. This survey, commissioned by the State of Alaska, concluded that collective damages totaled $2.8 billion. Federal Regulators are currently under order from Congress and the courts to determine ways and standards for evaluating losses suffered from environmental damage. Rather than supporting CV, however, this indicates a serious need for valuation methods of ecosystems services and attributes beyond the traditionally considered clean-up costs and impacts on measurable economic activity.

SUMMARY

In this chapter we discussed the three main types of market failure, intervention failure, and existence failure. In real-world situations there is usually no clear-cut distinction between the types of market and policy failures discussed in this chapter. There are elements of "public goods" in all environmental amenities. For example, the energy sector is characterized by natural monopolies and is also the largest generator of environmental externalities despite the fact that externalities abound in virtually every sector of the economy.

In light of the pervasiveness of these problems, the economist Richard Norgaard has argued for methodological pluralism in environmental economics. There can be no "unified field theory" in economics that will give us an unambiguous policy solution to every type of environmental problem. It is argued throughout this book that there are limits to the applicability of neoclassical theory to environmental problems. The starting point for standard theory is the individual, self-interest-oriented consumer. In many cases, such as biodiversity loss or global climate change, it is inappropriate to base public policy on individual preferences. There are other situations, however, when an individual-based approach might be justified. Furthermore, with almost all environmental problems, "internalizing the externalities" would move us closer to a solution.

One way to approach environmental policy issues is to proceed through various levels of analysis and policy choices. Consider the case of the spotted owl in the northwest United States. The first level of policy failure is intervention failure, including the direct and indirect subsidies for timber cutting in national forests resulting in prices that are too low and harvest rates that are too high. At the next level we have market failure in that the value individuals place on preserving species and ecosystems is not reflected in market prices. Preserving spotted owls gives positive utility to individuals that should be reflected in the market price for timber from old growth forests. Even within the context of neoclassical theory, therefore, actions can be taken that will result in a higher level of environmental protection. For example, remove intervention failures, internalize externalities, and evaluate the situation. Even though neoclassical economics has many limitations when applied uncritically to environmental policy, its arguments and policy recommendations can also make bad situations better. The economist Joan Robinson once remarked that although the market may not be a good master, it can be a useful servant.

Going beyond neoclassical theory are policies that recognize the implicit intergenerational inequities in market transactions and the uncertainties involved in calculating the value of the life support systems the biophysical world provides for us humans. Additional policies such as the safe minimum standard approach are called for to preserve vital and incalculably valuable resources.

This chapter concludes our presentation and critique of neoclassical theory. In the next chapter, we present a brief history of economic thought and discuss a new approach to the economy-environment relationship called ecological economics.

SUGGESTIONS FOR FURTHER READING

Bunker, Stephen. *Underdeveloping the Amazon*. Univ. of Chicago Press, Chicago, 1985.

Ciriacy-Wantrup, S.V. and R. Bishop. "Common Property as a Concept in Natural Resources Policy," *Natural Resources Journal* (October 1975), 713–727.

Daly, Herman. Review of *Free Market Environmentalism* by Terry Anderson and Donald Leal in *Ecological Economics 7* (1993), 173–187.

Galbraith, John Kenneth. *Economics and the Public Purpose*. Andre Deutsch, London, 1974.

Hardin, Garret. "The Tragedy of the Commons," *Science* 162 (1968), 1243–1248.

Norgaard, Richard. "The Case for Methodological Pluralism," *Ecological Economics* 1 (1989), 37–57.

Pearce, David and Kerry Turner. *Economics of Natural Resources and the Environment*. Johns Hopkins Univ. Press, Baltimore, 1990.

Swaney, James. "Common Property, Reciprocity, and Community," *Journal of Economic Issues* 24 (June 1990), 451–462.

7
FROM SUPPLY AND
DEMAND TO SOCIAL
AND ECOLOGICAL
CONTEXT

INTRODUCTION

This chapter discusses the larger issue of the social context of economic theory and policy and provides a short history of economic thought. Economic theory does not originate in a vacuum. More than in perhaps any other field of study, concepts and worldviews in economics arise from the social and political *milieu* in which economic activity takes place. Great economists like Adam Smith, David Ricardo, Karl Marx, and John Maynard Keynes all formulated their economic doctrines at crucial points in history; during periods of rapid change from one set of social, political, and intellectual conditions to another. To understand economic theory, therefore, it is necessary to understand the social currents present at the time these theories were conceived and promoted. The American economist Wesley Mitchell expressed this idea eloquently in the 1940s:

Economists are prone to think their work is the outcome of a play of free intelligence over logically formulated problems. They may acknowledge that their ideas have been influenced by their reading and the teaching which they were wise enough to choose, but they seldom realize that their free intelligence has been molded by the circumstances in which they have grown up; that their minds are social products; that they cannot in any serious sense transcend their environment.[1]

Some schools of economic thought have recognized and made explicit the influence of the context in which they were formed. Other schools perceive themselves as independent of such outside influences or even as shaping the surrounding conditions. Figure 7.1 shows the various levels of context that are explicitly considered by the schools of economic thought discussed in this chapter.

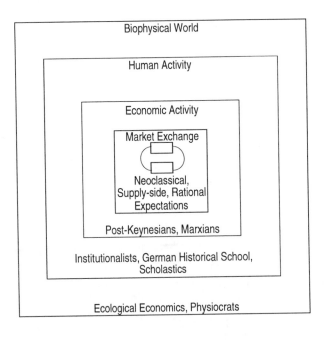

Figure 7.1 Schools of Economic Thought.

While some remain confined to market exchange with the rest of the world viewed as the support system for market activity, others go far beyond purely economic criteria. This chapter will highlight some of the main characteristics of these schools of thought, and once again focus on the implications of the evolution of economic theory for the biophysical world in which we live.

BEFORE THE CLASSICAL ECONOMISTS

Long before 1776 when Adam Smith wrote the *Wealth of Nations,* ideas about economic theory and policy abounded. In these early writings, however, economics was a part of larger concerns and not an isolated field. Three schools of early economic thought were the Scholastics, the Mercantilists, and the Physiocrats. The Scholastics were Medieval writers, and like most literate people of their day, churchmen. In contrast to the neoclassical view, which holds that questions of social welfare and ethics remain outside the focus of economics (see Chapter 4), for the Scholastics these concerns were inseparably linked. According to the Scholastics, principles of justice derived from Greek philosophy, Roman law, or the Bible should govern economic actions. Economic issues addressed by scholastic writers ranged from the distribution of income and wealth to a just price for the exchange of goods, and "usury" or interest rates. For example, Scholastics believed that the price charged for a good should not simply be established by market forces, but should be a "just price." To sell something for more than its just price or to buy something below it was considered unlawful. One way to establish a just price was to measure the value of a product by the amount of labor required to make it. Thomas Aquinas (1225–74) acknowledged another important influence on price determination—namely, human needs, because need is related to the value of a good. So is it just to sell at a higher price to a person in greater need? For the Scholastics, the answer was yes, provided that the price is set in a market free of fraud and greed. Others argued that a just price is one that allows buyers and sellers to *maintain* their social status, no more and no less. This points to the social context in which these economic considerations evolved. Medieval society was shaped by the strict hierarchical order of the feudal system that assigned to Lords, merchants, and peasants a particular place in society as determined by "natural law." Nature, too, was viewed as functioning according to a natural law outside of human control, but asserting its power in famine and disease.

With the dawn of the Enlightenment and a shift from a fixed, God-given order to human influence and control, Mercantilism emerged as a progressive force in its day. Mercantilists believed that a country's economic success was rooted in its accumulation of gold and silver. To ensure the accumulation of these precious metals, trade restrictions were imposed by the government. Thus economics became a matter of competition between nations in which each nation sought to outdo others by achieving a favorable balance of trade. Replacing the feudal system, strong central governments that promoted powerful trade companies to accomplish national interests of power and accumulation became the order of the day. Along with international trade, Mercantilism exported Euro-centered ideas of economic development that has had lasting effects on virtually every continent, their people, social organization, land, and natural resources.

A natural order of a different kind guided the Physiocrats' ideas of economic activity. They viewed their economic theory as innately evident (a doctrine of natural order) in a perfectly constituted society. The code of justice guiding such a perfectly constituted society included private property rights, freedom of contract, and freedom of trade. Thus the Physiocrats opposed Mercantilism and shifted the economic focus from national governments toward individual economic actors. The Physiocrats' most famous model is the circular flow model of goods and money. This concept was developed by Francois Quesnay (1694–1774), a French surgeon who took up economics in the 1750s and compared the flow of goods and money to the bloodstream flowing through the body. Quesnay's *Tableau Economique,* showing the circular flow of production among social classes, foreshadowed Wassily Leontief's input-output model that earned him the Nobel Prize in economics some 200 years later. For the Physiocrats, agriculture was the most crucial sector in the economy. Land could do something no other productive factor could, namely, produce rather than transform materials. Land was viewed as innately productive and distinct from human-made capital inputs. The health of the entire economy depends, therefore, on the health of agriculture, the only sector that can produce net surplus over the cost of production. But rather than translating this idea into a notion of sustainability and responsibility for maintaining nature's productivity, Physiocrats saw the health of the economy as dependent on employing sufficient capital in agriculture and removing Mercantilist restrictions on farming.

THE CLASSICAL ECONOMISTS

In the late 1700s Adam Smith formulated his well-known doctrine of the invisible hand. According to this doctrine, it is the activity of individual actors pursuing their own best self-interest that will also, "as if guided by an invisible hand," achieve the best interest of the whole society. To the rising middle class of merchants and entrepreneurs, the heavy hand of governmental restrictions on behalf of the large trading companies, restrictions that helped propel England to the status of a premier world power, stood in the way of further economic development. Policies that were once eminently successful had outlived their day.

The emphasis on the importance of the individual's influence on the whole parallels the Enlightenment's acknowledgment of human control over nature that shaped Adam Smith's world and continues to influence our understanding of science, progress, and development. Nature's bounty exists to be taken and controlled for human use. Francis Bacon effectively made the point:

I am come in very truth leading to you nature with all her children to bind her to your service and make her your slave. ...We have no right, to expect nature to come to us. Rather she must be taken by the forelock, being bald behind.[2]

No longer was the individual bound by a given social or even a natural order, but rather it was the individual whose impact shaped the whole. The unique quality of land (nature's input) still formed the basis for Smith's flow component of capital consisting of those inputs that "feed" production. But it was quickly replaced by a shift in focus from the value-creating properties of land (nature) to the value-creating properties of labor (humans), and from physical to monetary measures of output.

Adam Smith's spirited attack on the government regulations of the Mercantilists, although echoed by today's conservative opposition to social and environmental regulations, was not a free license for market forces to replace all rules. Smith's main premise was sympathy, that is, the capacity for human beings to feel for others. For Smith, this sympathy guided society's rules of ethics and justice. In one of his major works, *The Theory of Moral Sentiments*, Smith warned of the dangers of selfishness and acquisitiveness:

> The disposition to admire, and almost to worship, the rich and powerful, and to despise, or at least to neglect, persons of poor and mean condition, though necessary both to establish and to maintain the distinction of ranks and the order of society, is, at the same time, the great and most universal cause of the corruption of our moral sentiments....[3]

Recognized as the first great economist of the Classical school, Smith is often considered the originator of the major doctrines of modern neoclassical economics: the role of government should be minimal, and individual self-interest is the basis for a harmonious society. His writings, however, show a much more complex mind. Smith appreciated the need for government involvement in such things as public works for the betterment of society, and the need to care for society's less fortunate.

The French economist Jean Baptise Say (1767–1832) added another important component to classical economics and in fact, to the Physiocrats' flow diagram. He formulated *Say's Law*, which states that through the process of producing goods and services, inputs are employed (workers, owners of raw materials, suppliers of capital) who will receive exactly enough payments for their services to buy back the output produced. In other words, supply creates its own demand, and the flow of money in the economy is equal to the flow of goods and services. Say believed that more money available meant more money actually spent, and thus the circular flow was closed.

Within fifty years after Adam Smith wrote his great works in the late 1700s, the Industrial Revolution in Europe was in full swing. The social order changed dramatically. Three broad classes with different interests emerged, each with its own economic theorist as advocate. The interests of the landlords are associated with Thomas Malthus (1766–1834), those of the rising entrepreneurial class were championed by David Ricardo (1772–1823), and those of the growing working class by Karl Marx (1818–1883).

Malthus was a pastor before he specialized in economics and became the first professor of political economy in England. While his theory of population became a cornerstone of classical economics, his views on the theory of value and on Say's law were anticlassical, and are in many ways reminiscent of the Physiocrats. He asserted, for example, that commerce is inferior to agriculture as a source of wealth since its prosperity is only temporary, while agriculture's productivity is more permanent. His theory of underconsumption preceded that of Keynes by over 100 years . Malthus' well-known theory of population states that humankind will inevitably face misery, since population growth outpaces the growth of food supply. He wrote:

> It may be safely asserted, therefore, that population, when unchecked, increases in a geometrical progression of such a nature as to double itself every twenty-five years... If, setting out from a tolerably well peopled country such as England, France, Italy, or Germany, we were to suppose that, by great attention to agriculture, its produce could be permanently increased every twenty-five years by a quantity equal to that which it at present produces, it would be allowing a rate of increase decidedly beyond any probability of realization... Yet this would be an arithmetical progression, and would fall short, beyond all comparison, of the natural increase of population in a geometrical progression.[4]

Malthus' social views were harsh. He believed that poverty was a just punishment for the failure of the lower classes to limit their passion, and many of his ideas were incorporated into the repressive Poor Laws passed by the British Parliament in 1834. His views on the consequences of overpopulation are still ridiculed by those who reject the idea that humans must obey the rules of the biological world and see technology as the solution to all problems. Yet the ideas of Malthus have also received renewed attention, first in response to the Club of Rome's warnings and the "limits to growth" debate it sparked in the late 1960s, and more recently as a reaction to the return of the specter of starvation, especially in sub-Saharan Africa during the 1980s and 1990s. Many of these disasters are the result of social and political upheavals that have their roots at least partly in the inheritance of Mercantilism and colonialism, but they are also the consequence of human-created ecological disasters caused in the name of progress and development. Those who warn of the limits to economic activity are often referred to as "neo-Malthusians."

David Ricardo was a successful banker and stock-market speculator who made enough money from his investments to retire at the age of 43. He is recognized as one of the first deductive economic theorists. Like present-day economists, he began by constructing a set of basic premises and then logically deriving a set of consequences. Ricardo's friend Malthus, with whom he corresponded on many economic issues, considered him far too abstract. Ricardo, as a member of Parliament, successfully attacked the Corn Laws that protected English grain prices through tariffs on grain imports. These protections benefited landowners at the expense of employers who had to pay higher wages because of artificially high food prices. For Ricardo, it was the cost of production that ultimately regulated the price of a commodity. With Ricardo, the measure of the value of production also shifted to another level of abstraction. Malthus initially advocated a measure of value based on the ratio of the price of corn to the price of labor, and later a measure based on labor prices only. Ricardo proposed a measure based on the labor/capital ratio. Thus the move from concrete physical units of output to an abstract measure of value of output, and from land to labor and capital as sources of wealth creation, was complete.

The consequence of these abstractions is evident in Ricardo's best known theory, the theory of comparative advantage. The idea is that even if one country is absolutely more productive at producing all goods (both beef and Brazil nuts) than another country, it is still beneficial to both countries to specialize and trade with each other. Ricardo's argument is based on the notion of opportunity cost, which goes back to the distinction between economic and accounting costs introduced in Chapter 5. Opportunity costs include the cost of production alternatives that remain unrealized and thus represent foregone opportunities. If beef can be produced at twice the cost of Brazil nuts in country A, while in country B the ratio of Brazil nut to beef production costs is 1.5, then country B should produce beef and country A should produce Brazil nuts since the opportunity costs of beef production (in terms of Brazil nut production) are lower for country B than for A. This argument is still used today by advocates of free trade. Usually overlooked is Ricardo's warning that the theory breaks down if capital is mobile between the two countries. In that case, trade may be detrimental to one of the countries.

Karl Marx was a prodigious scholar who read widely and deeply in philosophy, history, natural science, and economics. He was influenced by Smith and Ricardo but recast their theories within his own unique framework. According to Marx's labor theory of value, the "exchange value," or what might be called the value in market trade, of a commodity depends upon the socially necessary labor time needed to produce it. If labor is the source of all exchange value, then it follows that labor should receive all the net product, or surplus value, of production.

Marx's theory of value has been misinterpreted by legions of economists who fail to see his distinction between "exchange value" and "use value," and between value and wealth. In a reply to a socialist manifesto, *The Gotha Program*, which asserted that labor is the source of all wealth, Marx wrote:

> Labor is not the source of all wealth. Nature is just as much the source of use values (and it is surely of such that material wealth consists!) as labour, which is itself only the manifestation of a force of nature, human labor power.[5]

Although Marx described natural resources as "free gifts of nature," what he meant by "free" was their effect on exchange value. He did not say that natural resources are unlimited. However, Marx vehemently disagreed with Malthus, since he believed that poverty was not the result of overpopulation but the result of the capitalist exploitation of the workers' labor. Marx's colleague Engels put his trust in science, which he believed to progress at least as fast as population.

One of the most interesting Classical economists, and one whose ideas are still relevant to today's "limits to growth" debate, is John Stuart Mill (1806–1873). Mill was the first great synthesizer of economic thought and wrote the first true economic textbook, *Principles of Political Economy*, which was widely used for over 40 years. The son of the philosopher, historian, and political economist James Mill, John Stuart published scholarly papers by the time he was 19. He made many contributions to economics and social philosophy. His 1869 paper "Subjection of Women" championed equal rights for women, arguing that the emancipation of women was necessary for the emancipation of men. He was a vigorous defender of freedom of the press, the rights of workers to form unions, birth control, and other "radical" reforms of the time.

Mill recognized that economic growth and accumulation must someday come to an end and that society must move from what he called the "progressive state" to a "stationary" state. Mill felt that this would be a positive development for humankind. He wrote:

> I cannot...regard the stationary state of capital and wealth with the unaffected aversion so generally manifested towards it by political economists of the old school. I am inclined to believe that it would be, on the whole, a very considerable improvement on our present condition. I confess I am not charmed with the ideal of life held out by those who think the normal state of human beings is that of struggling to get on; that the trampling, crushing, elbowing, and treading on each other's heels which form the existing type of social life, are the most desirable lot of human kind, or anything but the disagreeable symptoms of one of the phases of industrial progress.[6]

Mill thus questioned the notion of "more is better," one of the foundations of neoclassical utility theory as described in Chapters 2 and 5.

The Classical economists laid the foundations for modern economics. Despite the varied strands of influence in this school of thought, what is condensed from their message in modern neoclassical economics is the virtue of selfishness, the benefits of a limited role for government, and the idea that the economic system is independent of the natural world. The move toward an increasingly homogenous measure of money and output as value-creation allowed for a simple aggregation of all output into total product and all demand into aggregate demand. A closer look at the Classical economists, however, shows that they had a much more complex and qualified stance on both the social and ecological responsibilities of economic policy.

THE MARGINALIST REVOLUTION AND THE MATHEMATICAL FOUNDATIONS OF NEOCLASSICAL ECONOMICS

In the 1870s, a number of economists cast the ideas of Classical economics into the language of differential calculus. By this time both the scientific and industrial revolutions were in full swing. Trust in the human capacity for progress and betterment through science and technology had not yet been shaken by the events of World War I. This was the backdrop to the so-called "marginalist revolution" which, more than any other development in economic thought, shaped modern neoclassical economics. The marginalist revolution moved the focus of economic theory even further away from the environmental and social context within which economic activity takes place, to a cause-and-effect focus on small changes in prices and quantities. To accurately calculate the impact of small (marginal) change, the surrounding conditions must be assumed to be unchanged. Surrounding conditions simply form the constant, controlled environment in which the experiment takes place, a notion economics clearly borrowed from science.

Although many of the ideas of the marginalist revolution were developed by Heinrich Gossen (1810–1858) in the 1850s, Gossen's work remained unknown until it was discovered decades later. The main concepts of marginal analysis were discovered more or less independently by three people in three different countries: William Jevons (1835–1882) in England, Léon Walras (1834–1910) in Switzerland, and Carl Menger (1840–1921) in Austria. Most of the concepts we looked at in earlier chapters such as the "equimarginal principle"($MU_X/P_X = MU_Y/P_Y$), or the notion of mutual benefit from exchange (the basis of Pareto optimality), can be traced to the contributions of the marginalists.

The mathematical abstraction of economics was continued by the second wave of Marginalists during the last part of the 19th and early 20th centuries. Francis Edgeworth (1845–1926) introduced indifference curves, contract curves, and the Edgeworth box diagrams of exchange presented in Chapters 2, 3 and 4. The American economist John Bates Clark (1847–1938) added the production side to marginal analysis with the notion of marginal product and his marginal productivity theory of distribution. This theory was summarized in Chapter 5 where we saw that, under conditions of perfect competition, factors of production are paid according to their added contribution to total output.

The second great synthesizer of economic theory after John Stuart Mill was Alfred Marshall (1842–1924). His textbook, *Principles of Economics*, was published in 1890 and was widely used until the 1930s. Most of the concepts found in contemporary economic texts and discussed in Chapters 2–5 can be found in Marshall's writings. Marshall institutionalized modern marginal analysis, the basic concepts of supply and demand, and perhaps most importantly, the notion of economic equilibrium resulting from the interaction of supply and demand in the market. Marshall was well aware that the validity of an economic theory based on small changes around an equilibrium point depended on a stable environment. If the economic climate is characterized by sudden change and "exogenous" factors (originating from outside the system), then marginal analysis alone cannot explain economic conditions. To support his assumption of a stable surrounding environment, Marshall referred to Charles Darwin's dictum *natura non facit saltum*—nature does not make sudden changes. The kind of economic analysis pioneered by Alfred Marshall is called *partial equilibrium* analysis. It examines the forces of supply and demand in a particular market (as in the beef or Brazil nut markets) provided that all other influences can be excluded *ceteris paribus*. Unfortunately, the assumption of marginal analysis that outside conditions are stable often gets lost in neoclassical analysis. Thus the misleading conclusion is drawn that changes in particular economic units are independent of what is happening in the rest of the economy, society, or the biophysical world.

The *general equilibrium* analysis explored in Chapter 4 examines the conditions for stability for all the markets in an economy simultaneously. This moves us from Marshall's equilibrium conditions of supply and demand in individual markets (microeconomics) to the equilibrium conditions for the economy at large (macroeconomics). The Swiss economist Léon Walras established the mathematical principles of general economic equilibrium, but it was Pareto who applied Walras' principles and established modern welfare economics. As noted in Chapter 4, to construct a social welfare function, we need to leave the strict framework of neoclassical analysis and establish a set of social or ecological rules that helps us pick a particular Pareto optimal production and distribution of goods. While reminiscent of the question raised by the Scholastics as to what constitutes just market conditions, Pareto optimality is not based on a notion of a natural social order but on a belief in the ability of individual self-interest to insure the greater social good.

Another cornerstone of neoclassical-based macroeconomics is the quantity theory of money described by the *quantity equation* PQ ≡ MV. This identity states that the money received from the sale of goods and services in the economy (PQ) must equal the amount of money spent in the economy (MV). P and Q stand for the price and for quantity of goods and services, M for the money supply, and V for velocity. Velocity is the number of times the money supply "turns over" in a year. It is calculated as a "residual"; if the value of goods and services sold in a year is $1 trillion, and the money supply is $250 billion, then the velocity has to be 4; that is, the average dollar changes hands (is spent) 4 times per year. The quantity equation formalizes what was stated in Chapter 5—that money in the neoclassical system is *neutral*. It is merely a means of exchange that has no effect on the real economy, the production of real output, or the employment level of real inputs. The function of money is simply to duplicate the results of direct barter and to assign comparable (relative) value to diverse goods and services. If we further assume that velocity (V) is constant, then the only effect of changing the money supply is to increase P, the price level, or the level of inflation.

By the late 1920s, economics had transformed the basic ideas of the Classical school, individual rationality as the basis of economic decision-making, and the tendency of the economy to achieve stability with minimal government involvement into a complex and self-contained system of thought—neoclassical economics. What is true for individual markets also holds for the economy as a whole;

it is a self-correcting and self-regulating system. According to this view, there can be no involuntary unemployment. If unemployment increases, the supply of labor is greater than demand, and wages will fall. As a result, employers would be enticed to hire more workers. Anything that interferes with this natural process such as labor unions or minimum wage laws would only do more harm than good. The comfortable world of neoclassical theory came to a crashing halt with the economic collapse that began with the stock market crash of 1929. Once again economic turmoil spawned a new economic theory.

THE BATTLE LINES ARE DRAWN: JOHN MAYNARD KEYNES AND *THE GENERAL THEORY*

With the worldwide depression of the 1930s, it become clear that the economy was not self-correcting. Unemployment in the industrialized world reached somewhere between one-fifth and one-quarter of the labor force, even though wages fell. This economic crisis lasted a full ten years, and only ended with the economic boom caused by World War II. The two basic notions of neoclassical economics, that the economy is self-correcting and that money is neutral, were the focus of a frontal attack by John Maynard Keynes in his great work, *The General Theory of Employment, Interest, and Money*. Keynes argued that an economy could settle into an *underemployment equilibrium*, a stable situation of less than full employment, which could persist for years.

Keynes attacked Say's Law reflected in the quantity theory of money, and introduced two ideas basic to his theory—uncertainty and the non-neutrality of money. Just because people received income in one time period, he argued, does not mean that they will spend it in the next time period. In a barter economy, Say's Law holds that one commodity (C), for example, beef, is directly traded for another, Brazil nuts (C'). Thus we get the transformation C → C'. When money enters the picture, Say's Law implies an uninterrupted flow of commodities, their transformation into money as a medium of exchange, and back to commodities as C → M → C'. Keynes argued that individuals do not necessarily spend the money they receive as compensation for their (input) services, and so this transformation may be interrupted. In times of uncertainty, investors may refrain from investing and consumers may refrain from consuming. The mere production of goods does not necessarily create demand in a smooth continuous process from one time period to another. Supply does not create its own demand.

Keynes argued that governments should play an active role in the management of the economy through *monetary policy* (that is changing the money supply) and through *fiscal policy* (that is changing taxing and spending). According to Keynes, the role of government was to smooth out the ups and downs of the economy and to stabilize economic growth by controlling aggregate demand through purchasing from the private sector, providing employment, and encouraging or discouraging private spending with changes in taxation. This completely changed the role of government in Western democracies. Once again a world crisis had a profound impact on the evolution of economic ideas as governments throughout Europe and the Americas adopted active economic stabilization policies.

With Keynes the focus of economics shifted not only from supply to demand, but continued the shift from the "real economy" of physical input and output to the economy of money and credit. What became decisive for economic growth and stability was not simply the provision of goods and services but the availability of savings and money supply to accommodate their production. Sufficient money supply and sufficient aggregate demand determine economic success. Writing as he did during the Great Depression, Keynes identified a successful economy as one with growing output and employment. This does not mean, however, that Keynes was uncritical of economic growth. In a letter to his grandchildren in 1932 he wrote:

> When the accumulation of wealth is no longer of high social importance, there will be great changes in the code of morals. We shall be able to rid ourselves of many of the pseudo-moral principles which have hag-ridden us for two hundred years, by which we have exalted some of the more distasteful of human qualities into the position of the highest virtues. We shall be able to afford to dare to assess the money-motive at its true value. The love of money as pos-

session—as distinguished from the love of money as a means to the enjoyments and realities of life—will be recognized for what it is, a sometimes disgusting morbidity, one of those semi-criminal, semi-pathological propensities which one hands over with a shudder to the specialists in mental disease.[7]

Today Keynesian policies are under unrelenting attack from supporters of a "hands-off" view of government. But the success of macroeconomic stabilization policies should not be forgotten. The fact that there has not been a major economic crisis in Western industrialized countries for over 50 years is a remarkable achievement considering the recurring crises during the first one hundred years or so of capitalism. The battle between Keynes and his critics is one that has been played out in many different settings for literally hundreds of years: the disagreements focus on whether there is a "natural" order in the world that works best if left alone, or whether there is a collective responsibility to make the world a better place.

THE NEOCLASSICAL SYNTHESIS

World War II brought an end to the Great Depression, and the economies of Europe and North America boomed. Unemployment in the United States dropped from double-digit levels throughout the 1930s to less than 1 percent during some of the war years. After the war the United States, having suffered no physical damage and having a country full of people with money to spend, became the world's dominant economic power. In the 1950s over half the world's economic output was produced in the United States. Likewise, the center of the mainstream of economic thought shifted from Great Britain to the United States. The third great synthesizer of economic thought after John Stuart Mill and Alfred Marshall was an American, Paul Samuelson, who published his classic work, *The Foundations of Economic Analysis,* in 1947. In this book Samuelson formalized and synthesized the mathematical neoclassical analysis of earlier generations and made it compatible with the macroeconomics of Keynes.

In addition to providing the third great synthesis of economic theory, Samuelson also completed the shift from natural inputs (land) to capital and labor alone, and from physical to abstract measures of capital. His theory of capital as being perfectly malleable (divisible into infinitely discrete units) sparked a lively debate with Joan Robinson of Cambridge, England (the so-called Cambridge controversy). Robinson argued that it is logically impossible to define a "price" for the input capital independent of the price of the product this capital is producing. By the early 1960s, however, Samuelson could justifiably say that the only economists who were not neoclassical were those of the extreme left and extreme right.

REVOLUTION AND COUNTERREVOLUTION: THE END OF CONSENSUS

The economic miracle of the 1950s and 1960s came to an abrupt end with the Arab-Israeli war of 1973 and the resulting Arab oil boycott. Real energy prices tripled between 1973 and 1980, a trend that seriously affected the world economy. A new economic term, *stagflation*, was introduced—meaning a simultaneous increase in unemployment and inflation. Compared to the robust economic growth of the postwar period (2.2 percent per year between 1950 and 1973), the annual rate of real GNP growth between 1973 and 1993 fell to only 1.2 percent. Income disparity both within and between countries increased as well. In the United States, the median family income peaked in the mid-1980s and has been declining since then. At the same time questions resurfaced about the limits of nonrenewable resources, the limits to economic growth, and the limits of the carrying capacity of our planet.

The effect of this turn of events was to open the door to various critics of the postwar neoclassical synthesis. When a system of belief is threatened there are two kinds of reactions—an inward turn toward superorthodoxy, and the rise of heretical alternatives to established beliefs. Both of these reactions occurred in the 1970s. The first attack came from the ultraconservatives and was led by the monetarist school, most notably by Milton Friedman. Monetarism is not really a separate school of thought but can be seen as more doctrinaire branch of neoclassical economics. Monetarists believe in the quantity equation of money ($PQ \equiv MV$) and like their Classical precursors, view the economy as self-correcting. They believe that government policies can have little effect on the "real" economy. Stimulation policies will merely lead to an increase in the level of inflation. The effects of stimulation policies on employment and output are also considered to be minimal since there is a "natural" rate of unemployment and output that cannot be affected by fiscal or monetary policies. The only role government should play, therefore, is to keep money supply (M) growing at the same rate as real (physical) output (Q). Assuming, as monetarists do, that velocity (V) is constant, this steady growth rate rule will keep the price level constant.

In recent years, monetarism has come under attack after the monetarist experiment in the late 1970s by Federal Reserve Chairman, Paul Volker, contributed to rising interest rates and a stagnating economy with high unemployment. The economic stagnation in Margaret Thatcher's England also showed the adverse effects of a tight money supply on real economic activity and the impossibility of controlling the money supply with any degree of accuracy.

Other superorthodox reactions came from the supply-side and rational expectations schools of thought. The rational expectations hypothesis (REH) takes the assumptions of perfect competition to their logical absurdity. According to this theory, producers and consumers not only have perfect information about the present, but can also, on the average, correctly predict the future. As a result, no government action can have a lasting effect, since the actors in the economy will always correctly anticipate government policies and take appropriate actions to offset it. Another school of thought, supply side economics, traces its roots back to Say's law and the belief that supply creates its own demand. Accordingly, the key to economic success is to encourage investment and production by giving more money to the owners of capital. Thus increased capital will increase investment, which will increase production, which in turn increases wage payments to workers and stimulates consumer demand. On the other side, welfare and unemployment insurance should be severely curtailed since these programs discourage people from working. As John Kenneth

Galbraith put it, supply-siders believe that the problem with the economy is that the poor have too much money and the rich do not have enough.

The economic slowdown that began with the oil price shocks of the 1970s also encouraged several "heretical" schools of economic thought. The post-Keynesians, led by Paul Davidson and the late Sidney Weintraub, believe that the ideas of Keynes have been grossly oversimplified and misinterpreted in the neoclassical synthesis. In particular, they object to the recasting of Keynes' idea of *pure uncertainty* as *risk*. Risk means that a particular probability can be assigned to an outcome. If a probability can be assigned, then data about economic events can be adjusted according to their probable occurrence, like the way we adjust for discounting based on current interest rates. Once these adjustments are made, marginal analysis can proceed as usual. What Keynes meant by uncertainty is something completely different from risk, as shown in the following from his 1937 article, "The Theory of Employment":

> By "uncertain" knowledge, let me explain, I do not mean merely to distinguish what is known for certain from what is only probable. The game of roulette is not subject, in this sense, to uncertainty; nor is the prospect of a Victory bond being drawn. Or, again, the expectation of life is only slightly uncertain. Even the weather is only moderately uncertain. The sense in which I am using the term is that in which the prospect of a European war is uncertain, or the price of copper and the rate of interest twenty years hence, or the obsolescence of a new invention, or the position of private wealth owners in the social system in 1970. About these matters there is no scientific basis on which to form any calculable probability whatever. We simply do not know.[8]

If pure uncertainty exists, then precise calculations of expected profit, expected discount rates, and the expected impacts of increases in private sector investments on unemployment and output are impossible. Post-Keynesians believe that the economy is inherently unstable and that government has an essential role to play in avoiding economic crises. They also reject the explanation of price formation based on marginal costs (see Chapter 5) and favor a *mark-up* pricing theory instead. In this view prices are derived from production costs plus some standard mark-up. Post-Keynesians also favor income policies, that is, measures to offset the tendency of capitalist market economies to generate income inequality. Although they are concerned with questions of justice and social welfare, post-Keynesians have had little to say about the impact of economic activity on the biophysical world. Their main concern, like Keynes, is to promote policies that encourage the growth of output and employment.

Another major alternative school of economic thought, with deep roots in the past, is institutional economics. The leaders of the American institutionalist school were strongly influenced by the German historical school led by Gustav Schmoller (1838–1917), who was critical of the deductive and abstract focus of classical economics. For Schmoller, no aspect of economics could be adequately understood without considering its geographical, cultural, social, political and ethical contexts. The German historical school, in contrast to the Classical school, saw a positive role for governments in promoting the common interest. American institutionalists include Thorstein Veblen (1857–1929), author of *The Theory of the Leisure Class*, who first used the term "neoclassical" to describe the economics of Alfred Marshall, and the economists Wesley Mitchell (1874–1948), Clarence Ayres (1892–1972), and John Kenneth Galbraith. Institutionalists take an evolutionary approach to the study of economics rather than viewing economic activity as in a state of general equilibrium. They emphasize the role of institutions in economic change and the importance of historical differences between individual countries, and advocate an active role for government in promoting the common good. They also emphasize, however, the role of technology in economic progress and generally reject the view that environmental problems are endemic to the industrial system. Instead, these problems are considered to be the result of the failure of institutions to apply the technology appropriate to the solution of ecological problems.

Marxian economics has a long history in the United States and has made important contributions. The Dean of American Marxians, Paul Sweezy, has published the journal *Monthly Review* since the early 1950s. Marxians not only emphasize the class conflict inherent in capitalism, but have also recast Marx's analysis in mathematical terms to show the importance of income distribution on the level and composition of employment and output. Like institutionalists and post-Keynesians, Marxians tend to equate progress with increased economic output without regard to the impact of such economic growth on the biophysical world. In the words of the late socialist Michael Harrington, "if abundance is not possible then neither is socialism." A notable exception is a new field within Marxian economics called "eco-socialism" that explores the connections between economic and environmental exploitation. Marxians have had a difficult time since the collapse of the planned economies in Eastern Europe, because, fair or unfair, people will forever equate Marxian economics with the failed collectivist policies of Eastern European communism.

While not an independent school of thought, advocates of input-output analysis have tended to be critical of neoclassical economics and what Leontief called "implicit theorizing." Input-output analysis is a method of economic analysis using a table showing the flow of economic activity through the economy. The column entries of an input-output table show the origin of inputs for a particular industry, for example, the various amounts of fertilizer, feed, energy, and steel used to produce hamburger beef patties. Across the table, its rows show where the output of each industry, (for example, beef and Brazil nut production), goes. Input-output analysis can show how changes in consumer demand in one industry affect (directly and indirectly) the output of all the other industries in the economy. It can also be used to calculate the direct and indirect use of energy and raw materials resulting from a change in output. Users of the input-output approach tend to favor practical descriptions of real world problems. Post-Keynesian, Marxian, and Ecological economics (see below) have made extensive use of input-output techniques, although its use is not limited to these schools. In contrast to neoclassical analysis, input-output analysis focuses on the structure of economies and economic activities.

NEW DIRECTIONS IN ECONOMIC THEORY AND POLICY: ECOLOGICAL ECONOMICS

If the oil crisis of the 1970s raised questions about the stability of economic systems, the global environmental crises of the 1960s and 1970s raised questions about the stability of ecological systems as well. The impact of economic activity on the biosphere, and the dependence of the economy on natural resources became increasingly clear. Public awareness of ecological disasters like the Waldsterben (the death of the forests) caused by acid rain or the impact of pesticides described in Rachel Carlson's *Silent Spring* added fuel to the discussion about the real cost of economic growth. This spawned a new school of economic thought—ecological economics. Although environmental concerns had been expressed earlier, most notably by the German economist William Kapp, it was Kenneth Boulding (1910–1992) and Nicholas Georgescu-Roegen (1906–1994) who were instrumental in founding a new school of economics, which explicitly acknowledged that the economic system is a subset of larger processes of the biophysical world. In his now classic article, "The Economics of the Coming Spaceship Earth," published in 1966, Boulding made the distinction between a "cowboy" economy that uses resources as if they are unlimited, and a "spaceship" economy that recognizes the limitations imposed by the natural world. Georgescu-Roegen, in his monumental work, *The Entropy Law and the Economic Process*, argued that the economy is not analogous to a reversible mechanical system, the model for Classical and neoclassical economics, but rather is subject to the second law of thermodynamics, the Entropy Law. Matter and energy are continually degraded by economic activity from usable (low entropy) to unusable (high entropy) forms. Georgescu-Roegen argued that the economy is not a mechanical, circular flow, but rather a one-way flow with natural resource streams entering and waste streams leaving the system. A finite environment requires that flows through the system be minimized.

Herman Daly, one of Georgescu-Roegen's students and the author of *Steady State Economics* and *Toward a Steady State Economy* extended and popularized many of these ideas. Daly argues that a growth economy cannot be sustained and that instead the focus should be on a steady state economy that seeks to maintain a sustainable level of economic activity. In addition to ecological sustainability, the steady state view addresses ethical concerns as well. Achieving or maintaining sustainable levels of output is not simply a matter of scale, but must address issues of distributional justice as well.

Ecological economics also criticizes standard measures of economic success such as the gross national product, or GNP. As mentioned in Chapter 1, GNP measures the total dollar value of the output of an economy in a given year. Economic activity that is not registered in monetary units, such as growing one's own food or contributing the support services of child rearing and homemaking, is not counted as a contribution to GNP. Such subsistence and support services are summarized under the rubric "informal sector." Likewise, the supportive contributions of ecological systems are not counted in economic performance indicators like GNP. However, if the neglect of subsistence or ecological functions creates costs in the form of environmental clean up expenses or added security expenses, this shows up as an increase in GNP, and thus indicates an improvement in economic performance. To address these erroneous signals, Daly and others suggest an "Index of Sustainable Economic Welfare," which adds such considerations as the degree of income inequality, public expenditures on health and education, energy consumption, or the loss of agricultural land.

A distinctive feature of ecological economics is its *transdisciplinary* character. Particular studies contributing to this approach may include principles of ecology, economics, geology, and atmospheric science. The emphasis of ecological economics is on sustainability, a rather ambiguous term which suggests that the goal of economic policy should be an economy compatible with the long-term requirements of a stable biophysical world. Unlike most neoclassical environmental economists, who seek to describe nature's services and functions using the framework of neoclassical economics, ecological economists strive for a true dialog with natural scientists. The distinction is made by Robert Costanza:

Environmental and resource economics, as it is currently practiced, covers only the application of neoclassical economics to environmental and resource problems. Ecology, as it is currently practiced, sometimes deals with human impacts on ecosystems, but the more common tendency is to stick to "natural" systems. *Ecological economics* aims to extend these modest areas of overlap. It will include neo-classical environmental economics and ecological impact studies as subsets, but will also encourage new ways of thinking about the linkages between ecological and economic systems.[9]

Policies based on ecological principles recognize the limitations of applying market rationality to decisions affecting ecosystems. The technological optimism of neoclassical theory is replaced with what Costanza calls "prudent pessimism." In making decisions affecting irreplaceable resources, especially the life support systems of the planet, we should err on the side of caution.

SUMMARY

This chapter summarized some of the main schools of eco-nomic thought and showed that economic theories and models do not arise in a vacuum but are formed by the political, social, and cultural currents of their time. This is evident throughout economic history from the Scholastic writers, whose understanding of eco-nomics was firmly embedded in the understanding of a natural law that guided society's social order and its relationship to nature, to the Physiocrats and the Classics who sought liberation from the repressiveness of "natural law" and who followed the Enlightenment's view of nature as something to be explored and controlled. This was followed by an increasing mathematical re-finement of economic theory in the marginalist revolution and neoclassical synthesis, characterized by a movement away from recognizing the importance of nature to an increasing abstraction of production defined by technological progress and the value creating power of capital. The success of the marginalist revolution was only briefly interrupted in the 1930s when Keynes formulated his anti-neoclassical theories in response to the experience of the Great Depression. More recently the oil shocks of the 1970s, world-wide ecological disasters, and the demise of the planned economies of Eastern Europe have influenced economic thought and sparked both superorthodoxy in the form of monetarism and supply side economics, as well as alternative schools of thought like institutional, evolutionary, post-Keynesian and Marxian economics. While these alternative schools emphasize social context, the biophysical world remains largely outside their conceptual frameworks.

In contrast, ecological economics views economic activity as embedded in and influenced by process in the biophysical world. It has its roots in the ecological crises and the limits to growth debate of the 1960s and 1970s. Ecological economics is a new field still in its formative stage and contains elements from various economic schools of thought, as well as elements from biology, ecology, and other natural sciences. The next chapter in this book applies the principles discussed so far to the problem of groundwater contamination through intensive agricultural production.

NOTES

1. Wesley Mitchell, *Types of Economic Theory*, vol. 1, edited by Joseph Dorfman (New York: A.M. Kelley, 1967), 36–37.

2. Francis Bacon, quoted in Carolyn Merchant, *The Death of Nature: Women, Ecology and the Scientific Revolution* (San Francisco: Harper and Row, 1980), 170.

3. Adam Smith, *The Theory of Moral Sentiment*, 10th ed. (London: Straham and Preston, 1804 [1759]), 119.

4. Thomas Malthus, "A Summary View of the Principle of Population," in *Introduction to Malthus*, edited by D.V. Glass (London: Watts, 1953), 119; first written for *Encyclopedia Britannica*, 1830.

5. Karl Marx, "Critique of the Gotha Program," in *Marx and Engels: Basic Writings on Politics and Philosophy* (New York: Doubleday, 1959 [1875]), 112.

6. John Stuart Mill, *Principles of Political Economy*, 7th ed. (London: Longmans, Green and Co., 1896 [1848]), 453.

7. John Maynard Keynes, "The General Theory of Employment," in *The General Theory and After: Part II* (London: Macmillan, 1973), 369.

8. *Ibid.*

9. Robert Costanza, "What is Ecological Economics," *Ecological Economics* 1 (1989), 1.

SUGGESTIONS FOR FURTHER READING

Blaug, Mark. *Economic Theory in Retrospect*, 4th ed., Cambridge Univ. Press, Cambridge, U.K., 1985.

Boulding, Kenneth. "The Economics of the Coming Spaceship Earth," in *Toward A Steady State Economy*, edited by Herman Daly. W.H. Freeman, San Francisco, 1973.

Brue, Stanley. *The Evolution of Economic Thought*. Dryden Press, Orlando, Florida, 1993.

Carson, Rachael. *Silent Spring*. 25th Anniversary ed. Houghton Mifflin Co., Boston, 1987. (Originally published 1962).

Costanza, Robert. "What is Ecological Economics," *Ecological Economics* 1: (1989), pp. 1–7.

Duchin, Faye and Glen-Marie Lange. *The Future of the Environment*. Oxford Univ. Press, New York, 1994.

Galbraith, John Kenneth. *Economics and the Public Purpose*. Houghton Mifflin, Boston, 1973.

Georgescu-Roegen, Nicholas. *The Entropy Law and the Economic Process*. Harvard Univ. Press, Cambridge, Massachusetts, 1971.

Keynes, John Maynard. "The General Theory of Employment," in *The General Theory and After: Part II*. Macmillan, London, 1973 [1937].

———. "Economic Possibilities for our Grandchildren," in *Essays in Persuasion*. Harcourt-Brace, New York, 1932.

Malthus, Thomas. *An Essay on the Principle of Population.*

Martinez-Alier, Juan and Klaus Schluemann. *Ecological Economics*. Basil Blackwell, New York, 1987.

Marx, Karl. "Critique of the Gotha Program," in *Marx and Engels: Basic Writings on Politics and Philosophy*. Doubleday: New York, 1959 [1875].

Meadows, Donella H. and Dennis Meadows. *The Limits to Growth*. Potomac Association, New York, 1972.

Mies, Maria. *Ecofeminism*. Zed Books, London, 1993.

U.S. Department of Commerce. *Statistical Abstract of the United States*. U.S. Government Printing Office, Washington, D.C., annually.

Veblen, Thorstein. *The Theory of the Leisure Class*. Random House New York, 1934.

Waring, Marilyn. *If Women Counted: A New Feminist Economics*. Harper and Row, San Francisco, 1990.

See also the following journals: *The Journal of Post Keynesian Economics*, *The Journal of Economic Issues* (Institutionalist), *The Review of Radical Political Economics* (Marxian) and *Ecological Economics*.

8

THE CHALLENGE OF POLLUTION CONTROL: GROUNDWATER POLLUTION

INTRODUCTION

In this chapter we illustrate how the concepts discussed in this book impact the evaluation of environmental pollution and pollution control measures (see Figure 8.1). We use the example of groundwater pollution by nitrates. We choose this example not because it is the most serious pollution problem we face, but because it illustrates well the various ways in which economic concepts affect the valuation and protection of environmental resources. Virtually all the concepts described in Chapters 2 through 6 influence groundwater pollution valuation and policy in some way. Nitrate pollution of groundwater may not be as prominent an issue as ozone depletion, toxic waste, or global climate change, but it is nonetheless an issue of great concern. As groundwater tables are lowered and surface water quality deteriorates, groundwater pollution has become an issue of concern in many areas. Numerous other pollutants from landfills, or toxic waste sites may be equally as serious or more serious than the nitrate pollution problem discussed here. However, nitrate pollution is a widespread problem not

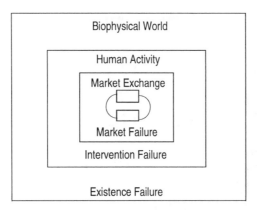

Figure 8.1 Links Between Production and Control Failures.

linked to local pollution sites, and it is one that illustrates well the links between production and pollution.

DETERMINING THE SOURCE OF POLLUTION

The first step in formulating pollution control policies is to identify the source of pollution. Nitrate groundwater pollution cannot be linked to a specific source like a smoke stack or a waste water pipe (point source pollution), but is caused by a broad based, non-point pollution. One identifiable source of groundwater pollution is organic and inorganic nitrogen fertilizer. There is not a simple cause and effect correlation between the intensity of fertilizer use and the level of groundwater pollution. Emissions vary with crop, soil type, precipitation levels and distribution, and hydro-geological conditions, to name just a few. One of the problems in correlating the level of fertilization intensity with the level of nitrate groundwater contamination is that emissions travel over space and over time. Nitrate levels can actually be reduced after emissions leave the pollution source but before they enter the groundwater layer itself. Not everything that leaves the plants' root zone also enters the groundwater. Soil and hydro-geological conditions play a role in these reduction processes or sink functions; the absorptive and regenerative functions of environmental media. However, nature's emission-reducing functions are not constant. They change over time, particularly as conditions suitable for nitrate reduction change with rising emission levels. The problem of linking source and effects is complicated by the fact that there are considerable time lags between the emission source and the environmental effect. Depending on soil and hydro-geological conditions, it may take 20 years or more for a nitrate problem to move through the soil into the groundwater layer and to accumulate so as to result in excessively high nitrate levels in the water pumped from the groundwater reservoir. The result is a threshold effect, that is, pollution effects are not noticed until a certain level of emission concentration is reached.

In policy analysis, these complex processes are generally reduced to identifying a main emission source such as intensive nitrogen fertilization. This cause-effect identification of pollution sources and pollution provides a starting point for taking appropriate policy measures. Policies are generally motivated by the need to reduce pollution in an environmental media (e.g., groundwater) but in order to reduce pollution, emission levels, that is, the pollutants leaving the pollution source need to be reduced. The focus on emission reduction in environmental policy is due to the fact that it is rarely possible to influence the processes that take place between the point at which emissions leave the pollution source (cause) and the pollution problem they create in the environmental media itself (effect). As is evident in our example, however, the emission from fertilization is only one factor among the many and varied influences that affect nitrate groundwater pollution. The difficulty here is to not let the complexity of processes in natural systems become so overwhelming as to impede corrective policy steps; on the other hand, policy should not become so oversimplified that measures are ineffective or even lead to detrimental results.

THE EFFICIENCY STANDARD

Once we have linked the pollution to its source, we are on our way to identifying marginal net private benefit (MNPB) and marginal social costs (MSC). MNPBs are based on the value of agricultural output. This output, in turn, is a function of nitrogen fertilization and a number of other input factors. So how do producers decide on how much fertilizer to use? To answer that question we can go back to Chapters 3 and 5 where we established the conditions for Pareto optimality in production. Consider production using three inputs, fertilizer (N), labor (L), and machines (K). We saw that resources are allocated optimally when the equi-price rule is met and the marginal product of fertilizer relative to its price is equal to the marginal product of labor relative to its price and to the marginal product of machine hours relative to its price. Or we can say: $(\Delta Q/\Delta N) \div P_N = (\Delta Q/\Delta L) \div P_L = (\Delta Q/\Delta K) \div P_K$. This can be written as $MP_N/P_N = MP_L/P_L = MP_K/P_K$.

This condition is not enough, however, to answer the question of how much fertilizer should producers use. Our rule for rational decision-making for producers was that producers seek to maximize profits. This was the case when the marginal costs of production (MC) are equal to the marginal revenue gained (MR). For our nitrogen fertilizer example, the additional (marginal) revenue a producer can gain from using an additional unit of fertilizer is the marginal product (or additional output, MP_N) gained from the extra input of fertilizer multiplied by the product price. This is called the *value of marginal product* or VMP. It is the marginal product ($MP_N = \Delta Q/\Delta N$) times the product price (P_Q). The additional cost of using more fertilizer is determined by the fertilizer price per unit. Thus we get the *value of the marginal product rule*, $VMP_N = P_N$.

Producers will add fertilizer (or any other input) until the additional revenue gained from its use (VMP_N) is just equal to the additional input cost (P_N). If the additional input cost exceeds the additional revenue generated, then expanding output does not make economic sense. To stop adding fertilizer at a level where the additional revenue generated from using more fertilizer is still higher than the additional fertilizer costs means using this input inefficiently. This is again assuming the conditions of perfect competition under which no producer can influence either the price of the product produced or the price of the inputs used in production. The case discussed here, agricultural production, is likely to be closer to the perfect competition ideal than most other markets.

So what is the result of this optimal input use condition? First, if the product price is high then the value of the marginal product of fertilizer use ($\Delta Q / \Delta N$) • P_Q will also be high. This means that even though each additional unit of fertilizer generates less and less additional output ($\Delta Q / \Delta N$ decreases), it may still be economically efficient to increase input use since the additional revenues gained are high (the output is worth a lot). The result may be fertilization intensities almost to the level of the plant physiological optimum—the point at which fertilization levels are so high that either the quality of the crop produced decreases or output decreases altogether with each increase in fertilizer inputs. Secondly, the lower the fertilizer prices, the lower the additional costs of increasing production. Even as the additional output gained from an additional unit of fertilizer (MP_N) goes toward zero, it will still be economically rational to increase fertilization levels if input prices are very low. This is the case for organic fertilizers where marginal input prices are close to zero. These kinds of fertilizers pose a waste problem rather than an optimal input use problem. Other kinds of rationality enter into production decisions.

This example also illustrates that the costs of groundwater pollution through nitrates are not part of the private benefit function. Private and social optima do not coincide. The costs of groundwater pollution represent negative externalities and do not enter the producers' (farmers') production decision. It is also important to remember that what we referred to as output is a collection of many agricultural products which are all produced using a common input (nitrogen fertilizer), but which have vastly different product prices, nutrient needs, and thus fertilization intensities, and production conditions (soil, hydro-geology, precipitation, etc.). The same monetary level of output of a particular production site (in this case, a farm) may have very different environmental effects depending on the products produced or the location of production.

MARKET PRICES VERSUS SHADOW PRICES

We assumed in our example above that the relevant prices in determining optimal production decisions are market prices. This allowed us to assume that the relevant benefit function we used to determine optimal pollution levels (see Figure 6.4) is based on marginal net private benefits (MNPB). But what if market prices are not optimal? This is the case, for example, when imperfect market structures exist, monopoly conditions being the extreme case (market failure), or when market prices are subsidized. In the case of nitrate pollution, the second scenario applies. Particularly in the European community, a vast number of agricultural products are subsidized. The purpose of these subsidies is to provide income support to farmers so as to keep family farms operating. While this may be positive from a social, cultural, or even land-use perspective, with respect to groundwater quality it is an example of *intervention failure*. One policy goal (income support to family farms) leads to a high value of marginal product (VMP) of fertilization and thus supports production intensities above the levels that would be considered optimal if product prices were not subsidized and thus lower. These lower product price levels that would exist if the ideal world of perfect competition prevailed, and no interference in markets took place, are called *shadow prices*. Estimates for selected products for the former West Germany show that fertilization intensities based on market prices are on average 11 percent higher than those based on shadow prices.

Our example illustrates the complexities of environmental policy decision-making. One socially desirable goal (income support to farmers) is in conflict with another socially desirable goal (groundwater protection). The problem is how to achieve the desired goal without any undesirable side effects. To address the problem of intervention failure, a rethinking of policy measures that were designed without an awareness of external consequences is required. This revised approach raises important institutional questions. The piecemeal approach to policy by which issues are considered separately by distinct institutions has failed in more areas than the one cited here. Viable solutions require serious coordination efforts of policy decisions crossing disciplinary and public agency lines. In particular, the coordination of ecological concerns, which are newer than the social concerns of employment, income, welfare, or education, require a rethinking of institutional structures and responsibilities. Frequently, administrative areas of expertise that have partitioned the structures of institutions into county, state, federal, or agricultural, economic, social welfare, and environmental compartments, are ill-equipped to respond to the complex linkages between human and ecological systems.

DETERMINING THE POLLUTION OPTIMUM

In the groundwater example, the neoclassical approach to environmental economics is to determine the optimal level of nitrate contamination of the groundwater. Since nitrate fertilization is the main source of nitrate emissions, the groundwater pollution problem can be examined as follows: the marginal net private benefits (MNPB) are measured as the benefits resulting from the production of agricultural products. The marginal social costs (MSC) are the costs associated with groundwater pollution, which is the externality of producing agricultural output.

We have already addressed some of the problems associated with measuring the MNPB function, such as market prices differing from shadow prices and output being an aggregate measure in which a variety of products and consequently a variety of emission levels are represented. The task of estimating the marginal social costs function is equally complex. In measuring the MSC of groundwater pollution by nitrates, one might consider direct and indirect health effects estimated by treatment costs, loss of working hours, the impact on social structures, or the drinking water supply of future generations, to name just a few. Or we might approximate this complex valuation problem by calculating the costs of correcting the damage done by fertilizer use—the costs of subsequent groundwater denitrification. The lower the allowed level of pollution, the higher the water treatment costs and vice versa.

By accepting subsequent denitrification as an approximation of the marginal social costs of nitrate contamination, we now have two different valuation frameworks. One uses nitrate contamination of the groundwater as a reference point; the other uses nitrogen fertilization levels. How do we answer the question, "What are the marginal social costs of nitrogen fertilization?" High nitrogen fertilization intensities also contribute to such problems as surface water eutrophication, changes in plant quality, or changes in biodiversity in areas surrounding highly intensive agricultural land, as some plants thrive under high nitrogen conditions while others are suppressed. All of these consequences of fertilizer use complicate the marginal costs assessment. To arrive at a manageable cost estimate one might argue that, since groundwater pollution is the issue at stake, accepting denitrification costs as an approximation of MSC is a legitimate proposition. In fact, such an approach isolates the impact of nitrogen fertilization on the groundwater emission problem, the very focus of our optimization problem.

This isolation does not take into account, however, two additional factors that complicate the problem of determining social cost. First, complementarity relationships exist among inputs. As nitrogen fertilization intensities are changed, so are the optimal intensities of other inputs, notably certain pesticides and phosphate fertilizers. The changing optimal levels of these other inputs mean that new marginal social cost functions which are the result of externalities caused by pesticides or phosphate fertilization would need to be constructed. These externalities would add to the marginal social costs of input use (nitrogen fertilization) but are unaffected if groundwater denitrification becomes the reference point of cost assessment. Secondly, accepting subsequent denitrification as a measure of marginal social costs raises the question of whose costs are considered. For example, what volume of water consumption do we base our estimate on—current drinking water use levels, future generation's prospective needs, or the entire amount of water estimated to leave the root zone? And how do we consider the impact of continually high fertilization intensities on the denitrification capability between the root zone and the groundwater, or within the groundwater layer itself, and what about the impact on microorganisms in the soil? These questions require the expansion of our usual human-centered perspective of evaluating groundwater pollution to include the impact on less obvious ecosystem functions. Some of these may be irreversibly harmed or harmed in ways that are only reversible in very long-term time frames. This illustrates the uncertainty involved in assessing the ecological impact of human economic activities. It involves more than assessing the risk of groundwater pollution alone. It is simply impossible to know all the processes that impact groundwater quality. Comparing prevention and clean up costs might be a permissible short cut to the complex valuation problem if it were not for the fact that, in some cases, it is simply impossible to clean up afterward. As the well-known example of pesticide accumulation in the food chain shows, prevention may well be the only alternative—clean up after the fact is impossible. Thus we can say that using denitrification costs as MSC estimates gives at best a minimum estimate. The actual social costs are likely to be considerably higher.

THE "SECOND BEST" APPROACH

The complex problem of determining a pollution optimum is usually side-stepped by imposing externally determined limits. Economists call this *the second best solution*. It is second best because it assumes that resources could be more optimally allocated by markets, but it is impossible to accurately determine the pollution optimum. Instead, an externally determined standard is accepted as setting the framework for resource allocation. In our example, the drinking water standard for nitrates recommended by the World Health Organization is 45ppm. This drinking water standard replaces the "pollution optimum." The resource allocation question thus becomes how can we best achieve this given drinking water standard, through reduction in fertilization intensity or through subsequent denitrification of the groundwater used for drinking water purposes? Estimates for selected products for German markets show that based on market prices, private fertilization optima exceed ecological optima (fertilization intensities which would lead to emission levels that do not exceed drinking water standards) by an average 26 percent. There are however, vast differences between products and locations ranging from no conflict between private and ecological fertilization optima to situations where ecological fertilization optima are exceeded by more than 50 percent.

The "second best" approach does not solve the problem of comparing the two policy measures—reduction of fertilization intensities and subsequent denitrification—which are not equal in their effect. As discussed earlier, pollution prevention is not the same as subsequent clean up. Subsequent denitrification must be viewed as a minimum condition for reaching the desired level of 45ppm nitrate emission, and the cost comparison of the two policies does not compare equal measures. As many examples show, the setting of standards does not end the resources allocation discussion. There are numerous examples of emission standards that were set for health or conservation reasons and were subsequently challenged since it was argued that they would be too costly to enforce. According to this argument, emission standards are more stringent than the optimal pollution level as determined by marginal net private benefits (MNPB) and marginal social cost (MSC) estimates. The debate about whether economics, natural science, or other considerations are to drive environmental policies remains unresolved. It is important to question, however, the assumptions upon which economic optima are based. Consideration of future use and ecosystem functions change economic estimates of pollution optima and are likely to result in more stringent standards. Attainment costs including option and existence value, may favor prevention measures that might be judged too costly if only present market values are considered.

THE POLICY QUESTION—HOW DO WE ENFORCE POLLUTION STANDARDS?

Once an optimal level or standard is determined, the question becomes what policies can be implemented to reach them? If denitrification turns out to be the least cost alternative, the question is answered relatively easily. The enforcement of drinking water standards would simply require water treatment at all sites where emission levels are exceeded. If emission prevention is determined to be the least cost measure, the policy question becomes much more complex.

The first policy option is to internalize the externalities and adjust prices to reflect the marginal social costs of nitrate emissions. This can be done either by lowering product prices, or, as in the case of subsidized products, by eliminating subsidies. Since lower product prices lower the value of marginal product of fertilizer use $(\Delta Q / \Delta N)$ • P_Q, optimal application levels would decrease. The effectiveness of this measure can be evaluated by estimating the product price elasticity of fertilizer demand,

$$\varepsilon_{PQ} = (\Delta N / \Delta P_Q)(P_Q / N),$$

which shows how input demand responds to changes in product price. Once the elasticity is known, we can calculate the price decrease necessary to reach desired reductions in nitrogen fertilization intensities. Even if the elasticities are such that lowering product prices promises to be an effective means of reducing fertilizer intensities, two points of caution must be raised. First, resulting income effects are likely to be severe, particularly to producers whose overall output levels are comparatively low. Other means of income support to compensate for decreased product price levels may be required. Second, fertilization intensities vary greatly with product and location. An across-the-board lowering of product prices may not achieve the ecologically desired result. While the production of some crops does not exceed emission levels even at current market price levels, others may far exceed them even when lower shadow prices are assumed. Crop price variations instead of across-the-board reductions are more likely to result in the production of a product mix that would meet desired pollution standards. Price policy measures also do not take into account regional variations that may be desirable from an ecological perspective.

Taxing nitrogen fertilizers would be another way to internalize externalities. To get an idea of the potential effectiveness of this method, one can calculate the price elasticity of fertilizer demand,

$$\varepsilon_{P_N} = (\Delta N / \Delta P_N)(P_N / N)$$

which shows how input demand responds to changes in input price.

This gives an estimate for the fertilizer price increases necessary to reach desired reductions of fertilization intensities. This assumes that producers indeed act as rational profit maximizers who base their use level on VMP = P_N. As with product price policies the disadvantage of this measure is that it may not allow for an ecologically desirable distinction between various crops and growing locations. In addition, fertilizer price increases do not affect emissions from organic fertilizers that may not be evaluated according to their nitrogen value at all, but may in some areas constitute a waste problem. However, the decisive point may not be whether price policy measures (input price increases or product price decreases) actually lead to the desired reduction in emission levels, but whether they lead in the right direction. Revenues from input taxes, for example, might be used to pursue other emission-reducing measures such as education and information about alternative emission reduction practices.

Education and moral suasion strategies can be used as alternatives to or in addition to price policies. Moral suasion refers to a strategy that seeks to convince producers to reduce production intensities or alter production methods not enticed by monetary incentives, but based on moral responsibility and ethical considerations. The basis for this approach is that there are in fact alternative production methods, which actually lower nitrogen emissions. Using three instead of two fertilizer applications, applying organic fertilizers in the spring rather than the fall, or planting a ground cover in the fall to lower percolation and nutrient loss during the winter months are just a few examples of how changes in production methods can effectively reduce emissions. Both in agriculture and industry, education about such measures has proven successful. Voluntary compliance is most promising when changes in production methods increase both social and private benefits, that is, reduce externalities and increase productivity. Production methods that are more in accordance with plant growth or precipitation pattern, for example, may reduce both emissions and production costs. Environmental management and eco-accounting focus on these areas of mutually beneficial changes from both an ecological and an economic perspective. Viable policies also include incentives for the implementation of production methods or products that are likely to bring about the desired reduction in emission levels.

Finally, regulations might seek to promote environmentally benign products, lower fertilization intensities, or alternative production methods to achieve lower emission levels. Such regulations could eliminate particularly problematic products, enforce a certain product mix, or limit production intensities. Such regulations, however, raise the question of enforcement. Unless education and moral suasion can establish that both social and private benefits result from regulations, enforcement costs are likely to be high.

SAFE MINIMUM STANDARD OPTIONS

Pollution problems such as the one discussed in this chapter challenge us to redefine neoclassical concepts based on a valuation framework of efficiency to one of sustainability. This does not mean that economic considerations are negligible but rather that questions of pollution reduction or resource protection need to be reevaluated within the larger context of ecological processes within which they exist. Considering the biophysical context, the optimization problem changes from one of allocating resources to reflect the market value of a resource to one that also reflects the maintaining and/or restoring of the sustaining qualities of environmental functions and processes. In fact, the neglect of such sustaining functions distorts the allocation problem, particularly in the long run. Take the example of nitrate reducing processes in the soil or in the groundwater layer. If these reduction capabilities are affected, then future fertilization levels that maintain emission levels below drinking water standards have to be lower than the ones that maintain drinking water quality under current conditions. Thus we have affected not just the water quality but also the productivity of future generations (displacement over time). The consideration of non-human ecosystems functions does benefit humans. But to consider what is obviously useful to humans may not be sufficient in determining ecologically sustainable emission levels. We may well be ignorant to the beneficial effects of seemingly unimportant microorganisms.

Considering these interconnections between human and ecosystem needs and between present and future generations, we are challenged to shift the focus of our policy options from the protection of individual resource functions (drinking water) to the protection of whole resources (groundwater reservoir) or even entire resource systems within an area (watersheds). Integrated policy options may include the designation of extensive drinking water protection areas in which land-use restrictions apply. This strategy may be followed particularly in areas with large groundwater reservoirs even if they are not currently used to meet drinking water needs. Another possible approach is to consider whole bioregions, such as a watershed, in determining resource use, production intensities, and production locations. Thus economic activities would no longer be considered place-less, but would take into account the biophysical conditions of a specific location. This approach may require compensation between regions for lost use alternatives and resulting losses in possible employment opportunities, since, particularly under current conditions, the sustainability of each individual region may not be feasible. However, such ecological exchange strategies have to be given careful consideration. To simply say we can under-pollute over here and over-pollute over there may not be a feasible approach. Resource conditions, such as size and quality of a groundwater reservoir, and the ability of bio-regions to cope with emission levels may differ greatly from region to region. Simple trade-offs are almost certainly impossible but rather warrant complex considerations of complex ecosystem functions. While this does not necessarily mean that each individual region has to be sustainable in and of itself, watershed or other bio-regional management considerations underscore the importance of being sensitive to the particular ecological characteristics and functions of a bio-region and its resulting land-use needs.

Sustainability rather than economic efficiency standards are not a call for conservation at any price or a return to nature without consideration of economic needs. They are, however, a plea for interdisciplinary cooperation. The often sadly insufficient ideas economists have about what determines ecological function and processes need to be corrected through the input of other disciplines. In our groundwater example, biologists, hydro-geologists, soil scientists, and agricultural management specialists need to be consulted. Beyond the circle of experts, however, the inclusion of local knowledge and of those affected is essential.

SUMMARY

The groundwater pollution example discussed in this chapter illustrates the daunting task of assessing externalities. Market failures resulting from the externalities of production are not easily measured or rectified. Even the task of determining least cost alternatives to achieve an externally set pollution standard is a considerable challenge. This challenge starts with assessing the cause and effect of pollution, continues with the difficulty in evaluating the marginal costs and benefits of pollution determined by current use values, and ends with the challenge of including future and non-human use alternatives. In our example, the valuation process was further complicated by the existence of intervention failure. Intervention failure was the result of price policies in the agricultural sector that were conceived as income support, but which increased production intensities and thus exacerbate negative externalities. This shows that in order to determine successful policies to reduce emissions, it is important that cooperation and coordination between commonly separate disciplines and agencies take place. The example of management alternatives that reduce emissions and increase the efficiency of input use also shows that externalities are not simply a public policy concern, but a management concern as well. In order to allocate resources sensibly, we need to include emission considerations in the production process itself instead of viewing them as external to economic valuation models.

This calls for the expansion of economic models to include the sustaining functions of ecosystems. Malthus' premonition of limits to human activity and growth may have to be restated. What poses limits to our economic activity is not, as we have thought, in the availability of resources for use as inputs, but the availability of sinks, that is, ecosystem functions to cope with our emissions. Consequently, we can either include considerations of sink capacities and sink quality protection in our economic considerations so as to not exceed absorption and regeneration capacities, or we can be forced to adjust by the collapse of the ecosystem itself. As many historic examples show, this means more than a decrease in environmental quality; it has serious economic consequences as well. A sustainable level and process of economic activity cannot be achieved unless ecological processes are considered to be part of economic activity itself through reciprocal processes instead of serving simply as its support system. If we neglect that, adjustments will simply lead to a displacement of negative externalities from one sink to another, from one media to another, and from one generator for another, just as replacements from one input to another have taken place throughout economic history.

SUGGESTED READINGS

Booth, Douglas. *Valuing Nature: The Decline and Preservation of Old-Growth Forests*. Rowman and Littlefield, Lanham, Maryland, 1993.

Hanley, Nick and Spash, Clive. *Cost-Benefit Analysis and the Environment*. Edgar Elgar Publishing, Vermont, United Kingdom, 1993.

Norgaard, Richard B. *Development Betrayed: The End of Progress and a Coevolutionary Revisioning of the Future*. Routledge, London, New York, 1994.

O'Hara, Sabine U. *Externe Effekte der Stickstoffdüngung—Probleme ihrer Bewertung und Ansäetze zu iher Verminderung aus ökonomischer und ökologischer Sicht. Schriften zur Umweltökonomie*: Vol. I. Kieler Wissenschaftsverlag Vauk, Kiel, West Germany, 1984.

Peet, John. *Energy and the Ecological Economics of Sustainability*. Island Press, Washington, D.C., 1992.

Ravetz, Jerome. *The Merger of Knowledge with Power. Essays in Critical Science*. Mansell Publishing, London, New York, 1990.

Sale, Kirkpatrik. *Human Scale*. Coward, McCann, and Geoghegan, New York, 1980.

The Conservation Foundation. *Groundwater Protection*. The Conservation Foundation Publishing, Washington, D.C., 1987.

Von Weizacker, Ernest U. and Jesinghaus, Jochen. *Ecological Tax Reform: A Policy Proposal for Sustainable Development*. Zed Books, New Jersey, London, 1992.

9

NEW DIRECTIONS
FOR ECONOMICS,
THE ECONOMY, AND
THE ENVIRONMENT

TAKING STOCK OF WHERE WE ARE

This book presents the basic world view of neoclassical economics, a view held by the vast majority of contemporary American and European economists who perceive the world through the lens of market exchange. Neoclassical theory concisely and eloquently describes the process of market exchange and derives a collection of behavioral rules that result in smoothly operating markets whose goal and end result is Pareto optimality. The market price system serves to replicate the ideal world of a face-to-face barter situation, provided that the conditions of perfect competition are met.

In reality, however, market exchange does not work smoothly. Neoclassical economics explains the shortcomings of the market with the theory of market failure. If the conditions of perfect competition are violated, and price signals are incorrect for reasons such as the existence of externalities or public goods, the market fails to achieve Pareto optimality. The core of neoclassical environmental policy is to correct erroneous price signals so as to "internalize the externalities."

We argue in this book that, even when market failures are corrected, the world of Pareto optimal efficiency does not guarantee that the vital functions of the biophysical world are sustained. In fact, the goal of Pareto optimal efficiency as the basis for economic policy may itself contribute to the unsustainable practices that assert themselves as the external, unaccounted for effects of market activities. Pareto optimality as a goal needs to be amended when the effects of optimally operating markets on the biophysical world are taken into account. Pareto optimality does not guarantee an optimal scale of economic activity or an ecologically sustainable use of resources.

There are numerous theoretical flaws of neoclassical economics that we have not addressed. One such example is the controversy surrounding "ordinal" and "cardinal" utility measures briefly mentioned in Chapter 4. Neoclassical theory asserts that only relative or ordinal utility levels are needed to establish Pareto optimality and that interpersonal cardinal or absolute utility levels are not necessary. However, cardinal utility notions sneak into the assumptions of market optimality through the better-off/worse-off measure of indifference levels and in the idea of diminishing marginal utility. Another unaddressed problem is the importance of duration and the influence of the relevant timeframe on both utility and production functions. For further exploration of these topics, we refer the reader to the suggested readings at the end of the chapters. Neoclassical economists will complain that the criticisms contained in this book have been addressed in the academic literature. The fact is, however, that almost all neoclassical policy recommendations fall back on the model of perfect competition and its assumptions. For example, the pitfalls of the free trade dogma are well-known by economists. These include the problems of increasing returns to scale, capital mobility, and of course, the distortions caused by unequal levels of environmental protection in different countries. Yet in their policy recommendations, neoclassical economists invariably fall back on the most simple model of perfect competition and comparative advantage.

This does not mean that the power of markets as organizing mechanisms of economic activity should be ignored. Nor does it imply that ways of organizing economic activity other than the market system have not had negative impacts on the quality of our natural environment. Human history in the recent and distant past is full of examples of the negative influence of human economic activity on the natural world. Neither the planned economies of Eastern Europe nor the economic practices of early non-industrial agricultural societies guaranteed the protection of environmental quality and the sustainable use of natural resources. We simply argue that markets do not automatically guarantee sustainability as a by-product of a freely operating price system and the allocation of value on the basis of individuals motivated by self interest. There are conflicts between individual and social goals, there are conflicts between short-term and long-term goals, and there are conflicts between human and ecological goals. Or are there really? The point is that by accepting the underlying assumptions of the model of perfect competition, conflicts are ignored instead of being made explicit and discussed openly through informed public discourse.

ECONOMIC DECISION-MAKING AND THE BIOPHYSICAL WORLD

To many if not most natural scientists, current crises like biodiversity loss, climate change, the depletion of the ozone layer, and many other environmental disasters are symptoms of an imbalance between the socio-economic system and the rest of the world. While it is true that the human imprint on the natural world is as old as human history, it is also clear that this imprint is currently several orders of magnitude greater than anything experienced in the past. Although there have been repeated episodes of devastation of local environments since the adoption of agriculture, it is only in recent decades that our imprints have reached global proportions. Certainly one reason for the profound effect of human activity on the natural world is the fact that there are so many of us. With population growth still increasing exponentially, many warn that we are fast approaching, or have already reached, the carrying capacity of our globe. Yet carrying capacity is a function of both numbers and impact. This does not in any way diminish the seriousness of population growth, but it does raise questions about how we address the problem of population growth. Apart from controlling numbers, we have to face the that fact because we have multiplied so much, we have an even larger responsibility to rethink our impact. The conflict between individual self-interest maximizers and ecosystems becomes all the more severe with more numerous self-interest maximizers. This increases the chance of interference both within social and biophysical systems.

As we stress throughout this book, one of the most serious consequences of the growing human impact on the natural world is the loss of biodiversity. Biological diversity is thought to affect the stability of ecosystems and the ability to cope with crises. In the 570-million-year history of complex life on planet Earth, there have been five major extinction events. The loss of biodiversity after these major extinction episodes, as measured by the loss of marine families, ranged from about 20 percent (about 65 million years ago) to more than 90 percent at the time of the "great dying" (250 million years ago). According to biologist E.O. Wilson and others, the loss in biodiversity caused by human activity since the Industrial Revolution alone is somewhere between 10 and 20 percent. If current trends continue, losses are likely to reach 50 percent by the end of the next century. After each of these events, it took between 20 and 100 million years for biodiversity to recover to previous levels, a length of time between 100 and 500 times longer than the 200,000-year history of Homo Sapiens. Within the lifetimes of most readers of this book, a sixth mass extinction will have occurred, one not caused by a meteor or a major volcanic explosion, but by human economic activity.

The greatest single cause of the loss of biodiversity is habitat destruction, that is the destruction of the web of organisms and functions which support individual species. The view of an individual species supported by various others within the ecosystem is foreign to the way markets view the world. Thus the example of biodiversity illustrates the conflicting frameworks of economics and ecology. Market decisions fail to account for the context of a species or the interconnections between resource quality and ecosystem functions. To use our example, the value of land used for beef production is measured according to its contribution to output. Yet long before output and the use value of land decreases, the diversity of grass varieties, microorganisms in the soil, or groundwater quality may be affected by intensive beef production methods. As long as yields are maintained, these changes go unnoticed by markets and are unimportant to land-use decisions.

Another obvious conflict is the relevant timeframe in markets and ecosystems. The biophysical world operates in tens of thousands and even millions of years. The timeframe relevant to market decisions is very short. Particularly where economic policy is concerned, two to four year election cycles are the frame of references, and for investors and dividend earners, performance timeframes of three months to one year are the rule.

Space or place is another conflict area. For ecosystems, place is of the essence. As illustrated by our groundwater pollution example, soil quality, hydrogeological conditions, regional precipitation rates, and size and location of groundwater reservoirs are essential features of an area's sink capacities and thus its ability to cope with pollution. These capacities are not simply transferable from one location to another. For global markets, place is increasingly irrelevant. Topography, location and function within a bioregion, or local ecological features do not enter into economic calculations except as simple functions of transportation costs or comparative advantage. Production is transferable, and the preferred location is where relative production costs are the lowest.

Economic and ecological systems also operate in different units. The unifying measure of market economies is money. Progress is measured in monetary units. In contrast biophysical systems operate in physical units such as energy, CO_2 absorption, or parts per million of nitrate contamination. The communication of effects and changes in dollar and cent amounts is an influential and powerful tool. However, it may also be very misleading and in fact, may mask serious changes in environmental quality or function.

These impediments show that "rational" decisions involving resource use made by individuals with finite lifespans at a specific point in time may be totally irrational for the human species.

Georgescu-Roegen's admonition to "love thy species as thyself" is vitally important not only for reasons of environmental conservation, but for human survival itself.

ECONOMIC DECISION-MAKING AND HUMAN CULTURE

Neoclassical theory is only one piece of the puzzle explaining human economic activity. Although there are numerous other schools of economics, the neoclassical school is so dominant that almost every practicing economist in the Western world follows its approach to describe reality and to set policy and research agendas. Moreover, neoclassical economics has had a global impact on how the proper goals of human activity are defined. Our theories shape our minds. As Karl Polanyi argued over 50 years ago, the relentless rationalization of the market sweeps aside cultural differences and reduces all human values to calculations of relative costs and benefits. People all over the world buy the same products, see the same kind of entertainment, and acquire the same tastes and prejudices. Biodiversity loss is not the only kind of homogenization promoted by the expanding global marketplace. The pursuit of economic development is a case in point. For years, progress was measured by increases in GNP. Only after years of criticism have development projects been amended by calls to pursue ecological sustainability, to maintain local social structures, and to rely on indigenous and locally reproducible technology.

The same economic forces that are reducing biodiversity are also reducing socio-diversity. By socio-diversity, we mean the different social and economic arrangements by which people have organized their societies, particularly the underlying assumptions, goals, values, and social behaviors guiding economic arrangements and processes. We believe that the very assumptions and the valuation concepts on which mainline economic theory is based lead to the successive loss of non-market economic systems and the adaptive ability of the worldwide market system itself. A variety of social scientists are warning that the increasing homogeneity of our world economy is making us particularly vulnerable to environmental and social disruption.

Yet it is easy for those who have everything to argue against a more-is-better framework. It is easy to argue for future preservation and frugality if one has a comfortable house, no lack of food, and even luxuries like cars, VCRs, and microwave ovens. The goal of ecological sustainability thus raises essential questions of distributive justice and about the values assigned to the sustaining contributions of human and natural systems. If frugality and sustainability are to be valued, the contributions of those who practice and teach us these values have to be considered as immensely valuable. And if the pursuit of communal or systemic, rather than individual interests, is to be valued, the contributions of those providing and maintaining communal support systems, care and nurture to nuclear and extended families, or of those who have acquired intimate knowledge of local ecosystems need to be assigned prestige and recognition instead of being relegated to the "informal" sector.

A mature ecosystem has been described as one in which all species enhance the survival of the others while maintaining themselves. Such ecosystems are characterized by coevolution and cooperation, as much as or more than by competition. These characteristics should also describe a mature economy. New economic theories are needed to describe and account for these coevolving and cooperative processes in economic systems.

DECISIONS UNDER UNCERTAINTY

Our example of biodiversity loss illustrates another characteristic of the global environmental problems we face. The value of biodiversity is unknown and unknowable. We cannot see into the future, and we are discovering daily that we really do not know much about the intricate workings of natural systems. As argued throughout this book, neoclassical theory alone is not sufficient to deal with such complex issues as intergenerational resource allocation or the links between distribution and pollution. Problems in applying neoclassical valuation methods arise from the dynamic nature of ecosystems, risks that are potentially enormous, and the complementarity of phenomena. Human interference in natural systems alters the processes within the system itself, which in turn changes the very conditions on which human decision-making was originally based. Under such conditions, decisions about resource allocation or production cannot simply be based on probability-based risk assessment, positive discount rates, or extrapolations from currently known use alternatives. There is growing criticism of economic assessment strategies from natural scientists who contend that commonly used methods such as the "maximum sustainable yield" of fisheries, have led to the consistent overuse of resources. Increasing criticism is also coming from social scientists who argue that problems like CO_2 accumulation, global climate change, or biodiversity loss cannot be assessed with in a rational choice framework since the effects of decisions are simply unknowable. What is needed are solutions based on strategies that stress caution, reversibility, and plausible hypotheses rather than the pretense of scientific accuracy. Some economists have arrived at similar conclusions. Newer safe minimum standard approaches seek to expand traditional cost-benefit-based valuation frameworks to include safety margins based on non-economic valuation concepts. The United Nations Conference on Environment and Development wrestled with this problem under the "precautionary principle." This principle states that we cannot use uncertainty as an excuse for inaction. Uncertainty demands caution, not procrastination.

The combination of high risk, high uncertainty, and emergent properties change the familiar conditions for decision-making. Funtowitz and Ravetz speak of an inversion of the traditional domination of "hard facts" over "soft values." When decision-making cannot rely on hard facts, it has to rely all the more on soft values, that is, ethical and moral criteria. Funtowitz and Ravetz call for a new "postnormal" science that would expand current disciplinary limits of natural or social sciences. In this expanded decision-making framework, economic considerations are only one of a multiple set of criteria to be considered. Neither economic indicators nor economic activity itself is per se a determinant of human welfare. Economic criteria or economic organizing mechanisms, therefore, cannot be permitted to place value on the entire world around us. Such single-mindedness results in the loss of biodiversity and socio-diversity, the loss of different ways of thinking and perceiving, organizing and doing. Such losses may well prove counterproductive and in fact, devastating as we seek to live sustainably, within the limits of our biophysical world. We may do well to heed Aldo Leopold, who warned that if we are unsure of what we are tearing apart, we should at least save all the pieces.

WHERE DO WE GO FROM HERE?

Global environmental crises such as biodiversity loss and global climate change do not lend themselves to the kind of solutions offered by neoclassical economic theory. This does not mean that traditional economic policy tools such as marketable permits or systems of taxes and subsidies cannot be useful as part of a larger strategy to solve these problems. But it does mean that we are not given neatly packaged solutions to our problems by relying on economic optimization models. Instead we need to rethink what markets can and should do, and not leave ourselves to their mercy. A pertinent question for environmental economists, then, is what does economics have to offer in dealing with these crises? We suggest the following four useful strategies in order of difficulty.

De-Mystifying Economics and Seeking Interdisciplinary Cooperation

Since the linkages between human socio-economic and biophysical systems require that existing valuation frameworks be expanded, the need for multidisciplinary cooperation becomes obvious. However, in order for dialogue and cooperation across traditional disciplinary lines to take place, we need to first de-mystify economics (as well as other disciplines). This means that the hidden assumptions and value judgements underlying economic concepts of environmental problems need to be made visible so that goals can be discussed and directions questioned and corrected. It is one of the ironies of unquestioned assumptions that during the past decade, there has been a growing dissatisfaction with neoclassical theory within the economics profession, while at the same time policy recommendations based on neoclassical theory have had a growing influence on policy-makers. For example, a recent study by economist William Nordhaus quantifying the costs and benefits of various strategies to slow global warming concluded that "rigid emissions or climate stabilization approaches would impose significant economic costs," leading to the recommendation of a wait-and-see approach to the problem. Nordhaus' estimates of the dollar costs and benefits of greenhouse polices have been widely accepted even by environmental groups. Among Nordhaus' assumptions is that utility can be described by a CES utility function. How many of the numerous people who have favorably quoted this study realize that the constant elasticity of substitution assumption implies that all goods are equally substitutable? In this framework, ski vacations, TV sets, a stable climate, and biodiversity are all on an equal footing and are all treated as market goods whose worth is to be judged solely by consumer preferences. It is certainly appropriate to consider neoclassical estimates of the "economic costs" (as measured by market prices, hedonic prices, or contingent valuation) of environmental policies as one of many criteria. However, the assumptions behind these estimates should be clearly spelled out in a manner that can be understood by non-economists.

The current efficiency debate is another case in point. Most of us affirm the need to use resources more efficiently, but it is questionable that the efficiency of the market described by the general equilibrium conditions of Pareto optimality is really what we mean. It may well be that efficiency should mean minimizing the physical flows of materials and not their costs, saving instead of substituting, and setting safe minimum standards instead of relying on markets to do our moral and ethical bidding.

Strategies and Structures Instead of Optima and Marginality

The most pressing global environmental crises we face involve non-marginal change, pervasive uncertainty, and perhaps most importantly, timescales stretching across many generations. A focus on marginal changes around a static equilibrium point is therefore woefully inadequate. Instead, "least cost," or better "least damage," strategies should be compared to meet the goals or targets established in a multidisciplinary valuation process. A viable strategy of least damage, however, implies that criteria other than economic factors need to be considered. What economists can contribute to the assessment process of attainment strategies is to study the structural impacts and changes such strategies are expected to have. Promising work has been done using input-output techniques to develop scenarios that show the impact of changes in economic activity on environmental quality or resource use, or the impact of environmental restrictions on economic activity. An essential feature of this approach is to develop natural resource accounts based on physical units and social accounting matrices that focus on the linkages between economic and socio-cultural systems.

From Homogeneity to Diversity

Whatever strategies we choose, they must account for diversity. Diversity is not a hindrance, but rather a key to resolving the conflict between human and natural systems. There is more than one strategy to address ecological problems, particularly if one considers the vastly different regional conditions that influence the impact of human economic activity on the biophysical world. Strategies that take regional conditions into account and that allow for the modification of policies need to be given serious consideration.

A second problem with diversity is harder to address. From a biological perspective, there is little question about the value of biodiversity. But from a socio-economic perspective, embracing diversity requires considerable rethinking. We have too often thought of diversity as somehow a hindrance to agreement, a cause of conflict, and a problem rather than a solution. Streamlining seems to be the answer, in turn, making everybody else like us. The problem with diversity, therefore, is not simply a problem of "different-ness," but rather of how we evaluate this "different-ness." An economic framework accepting and even supporting flexibility and responsiveness, and a differentiation of strategies that allows for regional ecological differences to be considered would go a long way toward accepting diversity. This also necessitates that we rethink the role of institutions. To effectively respond to biophysical systems, institutions should be cooperative rather than competitive, flexible rather than unresponsive. The market as a decentralized system has a definite place in rethinking institutional strategies, but care should be taken that markets preserve, not eliminate, diversity and decentralization.

Toward New Economic Models

A more ambitious but perhaps even more pressing need is to develop theoretical models that show the dynamic relationship between human activity and the environment. The human economy and the environment are self-organizing systems operating on conflicting principles. How is the economy able to continue to grow by degrading the environment, the ultimate source of its growth? Georgescu-Roegen has pointed out that economies grow by sucking low entropy from the environment and discharging it back into that environment as high entropy waste. Is there a way to model this process that would give us insights as to how to get off the growth treadmill without bringing our entire socio-economic system to a complete collapse? Promising efforts are underway to establish a postnormal scientific methodology to approach the problem of co-evolving systems. Work is also being done in political science describing the interactions between environmental degradation, resource scarcity, and social conflict. Caution, however, needs to be taken not to replace old with new "optimal" strategies, but to help in reshaping our way of thinking about economic success and progress. Doubts about our current path and how it has impacted our quality of life are evident all around us. Successful models and strategies need to lay bare the false dichotomy of a we/they relationship between humans and the biophysical world. The choice is not between an anthropocentric or an ecocentric worldview. We need to accept the reality of a biocentric world and the fact that humans are part of a living biosphere and dependent upon it. Economic models are needed which make explicit the mutuality, and reciprocity of economic, social, and ecological systems as illustrated in Figure 9.1.

In the current political climate, none of the above steps may be achieved easily. It is all the more important to recognize that economic descriptions lead to economic explanations and economic perceptions. With its strong influence of policy decisions, mainstream economics has influenced how we look at the world. Yet an economic system should be seen as a means, not as anend. Economies are there to serve larger societal goals. Despite all laissez-faire propaganda of neo-liberal market economics, there are countless positive examples of how we interfere into markets via price poli-

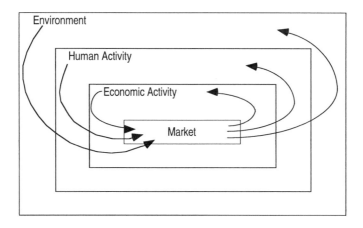

Figure 9.1 Contextual Valuation Framework.

cies and regulations in order to pursue policy goals established outside the economic systems. What we need to rethink is what directions our interference takes. We believe that economics still has important contributions to make both in identifying the goals and the means to achieve them. In order to become more relevant, however, economics has to clearly establish its place in the larger context of human affairs. A first step in this process should be an open debate about the hidden values our discipline brings to the policy discourse. We hope that this book has made a contribution toward an informed and open debate.

> Consider how future generations, whom we have left a destroyed earth, will despair when they learn of the unavoidability of their plight as they roam through old professional journals and learn that it was due to the fact that the calculation of integrals compelled their forefathers to consume more than the integrity of the planet would allow.[1]

1. Ulrich Hampike. "Neoklassik und Zeitpraeferenz: der Diskontierungsnebel," in *Die Oekologische Herausforderung fuer die Oekonomische Theorie,* edited by F. Becken Bach. (Metroplois, Marburg, 1991), 129.

SUGGESTED READING

Bormann, Herbert and Stephen Kellert. *Ecology, Economy, Ethics: The Broken Circle.* Yale Univ. Press, New Haven, 1991.

Eckersley, Robyn. *Environmentalism and Political Theory.* SUNY Press, Albany, New York 1992.

Funtowitz, Silvio and Ravetz, Jerry. "Science for the Post-Normal Age." *Futures* 25 (Sept. 1993).

Hern, Warren. "Why Are There so Many of Us? Description and Diagnosis of a Planetary Ecopathological Process," *Population and Environment* 12 (1990), 942.

McLaughlin, Andrew. *Regarding Nature: Industrialism and Deep Ecology.* SUNY Press, Albany, New York 1993.

Mies, Maria. *Patriarchy and Accumulation on a World Scale: Women in the International Division of Labour.* Zed Books, London. 1986.

Naess, Arne. *Ecology, Community and Lifestyle.* Cambridge Univ. Press, Cambridge, Massachusetts, 1989.

O'Hara, Sabine U. "Economic Reality, Gandian Ideas and Lessons from Socio-Diversity." *Ghandi Marg.* 16 (July/Sept.1994).

Sessions, George. *Deep Ecology for the 21st Century: Readings on the Philosophy and Practice of the New Environmentalism.* Shambhala Publications, Boston, Massachusetts, 1995.

Shepard, Paul. *The Tender Carnivore and the Sacred Game.* Scribners, New York, 1973.

Snyder, Gary. *The Practice of the Wild.* North Point Press, San Francisco, 1990.

Walker, James and James Kasting. "Effects of Fuel and Forest Conservation on Future Levels of Atmospheric Carbon Dioxide," *Palaeogeography, Palaeoclimatology, and Palaeoecology* 92 (1992), 151–185.

Wilson, E.O. *The Diversity of Life.* Harvard Univ. Press, Cambridge, 1992.

GLOSSARY

ACCOUNTING PROFIT—The profit as measured by accountants that is equal to sales (total revenue) minus accounting costs.

ADAPTIVE—Any characteristic that aids an organism or system to survive in its environment.

AIR QUALITY STANDARD—A prescribed level of pollution in the air that should not be exceeded.

ANTHROPOCENTRIC—Considering humans to be the central or most important part of the universe.

ASSIMILATIVE CAPACITY—The ability of the biophysical world, the air, water, and soil media, to absorb the waste products generated by economic activity.

BARTER ECONOMY—A simple exchange economy where consumers and producers barter directly with each other with no use of money or any other medium of exchange.

BIODIVERSITY—All of the species that currently exist on Earth, the variations that exist within each species, and all of the interactions that exist among these organisms and their biotic and abiotic environments, as well as the integrity of these interactions.

BIOPHYSICAL WORLD—The biological, geological, and atmospheric processes that make up the natural world upon which all economic activity depends.

CAPITAL—Durable produced goods that are used to produce other goods.

CARBON DIOXIDE—A gas, CO_2, making up about 3 percent of the earth's atmosphere. It is an end product of burning or oxidation of organic matter or carbon-containing substances.

CLASSICAL ECONOMISTS—The school of economic thought begun by Adam Smith in the 1700s. It included David Ricardo, Thomas Malthus, and John Stuart Mill. The basic view of the Classicals was that the economy is self-correcting and that the role of government should be minimal.

CLEAR CUTTING—The practice of removing all trees in a specific area.

COMMUNITY—Organisms or humans existing in a specific region.

COMPETITION—The struggle between individuals of the same or different species for food, space, mates, or limited resources. In economics, a term used generally to mean PERFECT COMPETITION.

COMPLEMENTARITY—In economics, referring to pairs of goods or productive inputs that are used together, such as hamburgers and hamburger buns. It may also refer to co-evolving systems that mutually influence each other.

CONSERVATION—The planned management of a natural resource to prevent over-exploitation, destruction, or neglect.

CONSUMER SURPLUS—The difference between the amount consumers actually pay for something and the amount they would be willing to pay.

CONSUMPTION—The act of purchasing, or trading for, goods and services exchanged in markets.

CONTRACT CURVE—A curve showing all the Pareto optimal distributions of goods (in consumer theory) or productive inputs (in production theory).

COST FUNCTION—An equation, schedule, or graph relating total costs of production per time period to the level of use of productive inputs.

DEMAND CURVE—A schedule or curve showing the relationship between the price of a good and the quantity demanded of that good, all other influences being held constant.

ECOLOGY—The study of the interrelationships between organisms and their environments.

ECOSYSTEM—The organisms of a specific area, together with the relationships between them and their functionally related environments.

EDGEWORTH BOX DIAGRAM—A diagram showing the exchange of goods between two consumers or the exchange of inputs between two firms. This diagram illustrates the logical underpinings of neoclassical theory and the notion of Pareto optimality.

ELASTICITY—A unit of measure widely used in economics to show the responsiveness of one economic variable to a small change in another economic variable.

ENVIRONMENT—The physical and biological aspects of a specific area.

EVOLUTION—A change in the gene frequency within a population; a change in a population's physical characteristics.

EXISTENCE FAILURE—The failure of an economic optimum to insure a biophysical optimum. Efficiency in market exchange will not guarantee the proper scale of economic activity in terms of ecological sustainability.

FERTILIZER—Any natural or artificial substance added to the soil to promote plant growth.

FOOD CHAIN—The sequence of organisms in a community, each of which uses the lower source as its energy supply. Green plants are the ultimate basis for the entire sequence in a food chain.

FOSSIL FUEL—Coal, oil, natural gas, and/or lignite, that is, those fuels derived from former living systems formed under extreme heat; commonly referred to as non-renewable fuels.

GENERAL EQUILIBRIUM THEORY—A situation in which all the markets in the economy are simultaneously in equilibrium. Equilibrium exists in a system when it will remain in its present state unless disturbed and, if disturbed, it will tend to return to its original condition.

GLOBAL CLIMATE CHANGE—The GREENHOUSE EFFECT and other worldwide changes in the composition of the atmosphere.

GREENHOUSE EFFECT—The effect noticed in greenhouses when shortwave solar radiation penetrates glass, is converted into heat (longer wavelengths) and blocked from escape by the window glass of the greenhouse. In the natural environment it refers to a similar effect whereby heat radiation is blocked from escape by Greenhouse Gases (for example CO_2, Methane) which accumulate in the atmosphere.

GROUND WATER—All water located below the earth's surface.

HABITAT—The natural environment of a plant or animal.

INDIFFERENCE CURVE—A curve showing the various combinations of the consumption of two goods that give a consumer the same amount of utility or satisfaction. The downward sloping indifference curve embodies the neoclassical assumption of the substitutability of any good with any other.

INTERVENTION FAILURE—A case in which government intervention in the private market moves the economy further away from a social optimum. Examples are government subsidies to environmentally destructive extractive industries.

ISOQUANT—A curve showing all the different combinations of two inputs that may be used to produce the same amount of output. Like the indifference curve, the isoquant embodies the notion of substitutability that lies at the heart of neoclassical analysis.

LAND—One of the three basic factors of production, along with labor and capital. In classical and neoclassical economics, "land" includes all natural resources.

LAND ETHIC—A term first used by Aldo Leopold to describe the philosophy of stewardship for the natural world.

LEACHING—Dissolving out of soluble materials by water percolating through the soil.

LONG RUN—A term used to describe that period of time in which all inputs are variable, including all capital equipment.

MALTHUSIAN THEORY—The theory that population tends to increase by geometric progression while food supplies increase arithmetically.

MARGINAL COST—The addition to total cost resulting from the production of one additional unit of output.

MARGINAL REVENUE—The addition to total revenue resulting from the sale of one additional unit of output.

MARGINAL PRODUCT—The addition to total product (total output) resulting from the employment of one additional unit of a productive input, the amount of all other inputs being held constant.

MARGINAL UTILITY—The addition to total utility gained from the consumption of one more unit of a good, the level of all other goods held constant.

MARKET STRUCTURE—The characteristics of the markets in which goods and services are bought and sold. Elements of market structure include the number of buyers and sellers, ease of entry and exit into and out of the market, and the characteristics of the product sold.

NATURAL MONOPOLY—A firm whose average production costs decline continuously as output increases. In this case, having only one firm means that the product can be produced more cheaply than having several firms.

NEOCLASSICAL SYNTHESIS—A term first used by Thorstein Veblen to describe the economics of Alfred Marshall, who cast the ideas of Adam Smith and the other Classical economists in the language of modern mathematics.

NITRATE—A salt of nitric acid. Nitrates are the major source of nitrogen for higher plants. Sodium nitrate and potassium nitrate are used as fertilizers.

OPPORTUNITY COST—The value of the next best use for the resources used to produce a good. The value of the best alternative to what is being produced.

ORGANIC—Derived from living systems.

OZONE—Molecule of oxygen containing three oxygen atoms; shields much of the earth from ultraviolet radiation.

PARETO OPTIMALITY—In consumption, a situation in which one person cannot be made better off without making someone else worse off, and in production, a situation in which the output of one good cannot be increased without decreasing the output of another good.

PERFECT COMPETITION—A type of market structure in which there are a very large number of buyers and sellers, there is perfect information about products, there is ease of entry into and out of markets, and the products sold in a given market are identical.

PESTICIDE—Any material used to kill pests such as mice, insects, bacteria, fungi, or other organisms considered pests.

PHOSPHATE—A phosphorous compound; used in medicine and as fertilizer.

POLLUTION—The process of contaminating air, water, or soil with materials that reduce the quality of the medium.

POPULATION—All members of a particular species occupying a specific area.

PRIMARY PRODUCTION—The energy accumulated and stored by plants through photosynthesis.

PRODUCTION FUNCTION—A graph, chart, equation, or table showing the relationship between output per time period and the use of productive inputs.

PRODUCTION POSSIBILITIES FRONTIER—A graph showing the menu of production possibilities of goods that can be produced by an economy when all resources are employed in the most efficient manner possible.

SAY'S LAW—The theory that supply creates its own demand, because paying productive inputs will create exactly enough money to buy back what is produced. Excess demand or supply is impossible.

SHORT RUN—A term used to describe that period of time in which some inputs are fixed, usually plant size and capital equipment.

SOCIAL WELFARE FUNCTION—A curve showing the points of social or societal well-being characterized by the utility levels of its members.

SOLID WASTE—Unwanted solid materials usually resulting from industrial processes.

SPECIES—Populations, or a population, capable of interbreeding and producing viable offspring.

SUPPLY CURVE—A schedule or curve showing the amount of a good firms will supply at various prices, all other things being equal.

TECHNOLOGY—Applied science, or the application of knowledge for practical purposes.

THRESHOLD EFFECT—The situation in which no effect is noticed, physiologically or psychologically, until a certain level of concentration is reached.

TOXIC—Poisonous; capable of harming living systems.

UTILITY—The total level of satisfaction derived from the consumption of goods and services.

UTILITY POSSIBILITIES FRONTIER—A frontier made up of all the points on the contract line signifying Pareto optimality in consumption. All points on the utility possibilities frontier show the Pareto optimal levels of utility consumers can achieve given an initial endowment with goods.

INDEX

DATE DUE